SpringerBriefs in Applied Sciences and Technology

Computational Intelligence

Series Editor

Janusz Kacprzyk, Systems Research Institute, Polish Academy of Sciences, Warsaw, Poland

SpringerBriefs in Computational Intelligence are a series of slim high-quality publications encompassing the entire spectrum of Computational Intelligence. Featuring compact volumes of 50 to 125 pages (approximately 20,000–45,000 words), Briefs are shorter than a conventional book but longer than a journal article. Thus Briefs serve as timely, concise tools for students, researchers, and professionals.

More information about this subseries at http://www.springer.com/series/10618

Soni Sweta

Modern Approach to Educational Data Mining and Its Applications

 Springer

Soni Sweta
Amity University Jharkhand
Ranchi, Jharkhand, India

ISSN 2191-530X   ISSN 2191-5318   (electronic)
SpringerBriefs in Applied Sciences and Technology
ISSN 2625-3704   ISSN 2625-3712   (electronic)
SpringerBriefs in Computational Intelligence
ISBN 978-981-33-4680-2   ISBN 978-981-33-4681-9   (eBook)
https://doi.org/10.1007/978-981-33-4681-9

This Springer imprint is published by the registered company Springer Nature Singapore Pte Ltd.
The registered company address is: 152 Beach Road, #21-01/04 Gateway East, Singapore 189721, Singapore

Your life was a blessing,
Your words were very inspiring…
This book is dedicated to my beloved father
Late Prof. (Dr.) C. D. Roy

Dr. Soni Sweta

Foreword

Being author of many books on Electronics and Communication Engineering, I found this book *Modern Approach to Educational Data Mining and Its Applications* is very relevant to the technology based educational system and all its stakeholders. What a nice deliberation in the domain of Computer Science and Technology and equally useful for all inter-disciplinary areas. Technology based learning enhances the proficiency in learning and overall efficacy of the education system.

The author of this book Dr. Soni Sweta has been associated with me since long. I encouraged and guided her achieving many milestones in the career and witnessed her successes through struggleful journey.

Dr. Soni Sweta is working as Assistant Professor in Department of Computer Science and Engineering in Amity University Jharkhand, Ranchi. She is a teacher, trainer, and consultant in the field of Information Technology. She is M.Tech. in Computer Science and Engineering from Rajiv Gandhi Proudyogiki Vishwavidyalaya Bhopal and Ph.D. in Computer Science and Engineering from Birla Institute of Technology, Mesra, Ranchi. Her research areas of interest include Artificial Intelligence, Data Mining, Soft Computing and Data Science. She has published many research papers in renowned International and National, SCI and Scopus indexed journals.

Education is very much essential for all human beings. It is desired by governments and academicians to impart education in cost effective and most adaptable ways to learners to upgrade learning skills. The author has stressed upon the evolution of education system from traditional to Artificial Intelligence inbuilt education system.

It is always challenging for teachers to fill the learning gaps while teaching. So, the technical aspects discussed in this book i.e. "Adaptive E-Learning Mechanism with Cognitive Approach" will certainly improvise stress free learning.

This book is divided into seven chapters and all chapters are closely knitted and very purpose full written in view of mitigating learning gaps and enhancing overall performance of any educational system. I am very much glad to discuss first two chapters based on educational data mining concepts and Techniques and Adaptive E-learning System. Chapter 3 i.e. the compilation of literature reviews strengthens the core ideas of this book and make this book incredible in terms of significance.

Remaining chapters give a wide view of the framework, implementation, and its cognitive applications in domain of education system.

Dr. Soni Sweta has more than 10 years of teaching experiences. So, I think she has enough knowledge in field of education system and her core subject areas. I appreciate her to put her experience and knowledge in form of book to give a large benefit to the society.

This book covers the multi-disciplinary area so it may be followed by multiple domains and the deliberations made by the author are equally important for all persons. I foresee adaptive e-learning system will be the new order of education system for effective learning in future. I think this is a good contribution by author in her early stage to the society. I wish her for future, new edition and many more books are awaited from her side.

Prof. (Dr.) D. K. Singh
Director
BIT Sindri
Dhanbad, Jharkhand, India

Preface

The Universe is a synonym of a lot of creations and activities. It is a dynamic world. Everything is changing in respect of others. The existence of life at heavenly bodies is one of the most important creations of the nature. It caused the genesis of numerous species across the world. Among them, human being is the most dominating one. They possess a high-skilled brain and adapt most suitably according to the changes in the surrounding environment and stimuli. They have a unique characteristic of sensing and learning. Their work starts from the womb, flourishes during the journey of life, and terminates at the tomb. Hence, learning is a lifelong process.

In fact, as time passes education became the core of the human development. It has promoted social, cultural, economic, and scientific temperament. It is a proven fact that learning enhances in the given conducive environment. Here, adaptivity plays a vital role and helps the learners to get the information most sought as per their own preferences and choices.

Many discoveries and researches have been carried out by applying herculean efforts to make a difference in way of thinking and living. Man-made machines help them in their day-to-day activities. As of now, they created a world of high-precision computers doing almost everything like human being without any fatigue and boredom. Man–machine collaboration has achieved its zenith. The concept of neural network, fuzzy logics, and fuzzy cognitive maps replicates the functioning of the human brain and nerves.

It is education which turned the wheel of human brain into the path of high-speed trajectory. It gathered momentum and became the most differentiating factor among them. There were a lot of experiments done in educational methodology to enrich the human soul. The brick-and-mortar system was at the center stage till recent past. Presently, in the advent of development of computers, information technology and availability of high-speed Internet have opened the floodgate of knowledge and education and made the existing educational system as the most redundant one. The way of traditional classroom teaching as teacher teaches students has changed significantly. The use of multimedia in the class has given different dimension of the education system. Gradually, it made the teaching interesting and improvised interaction levels among learners and teachers. But there are some constraints that are to be addressed well before achieving the desired harmonious and synchronous education system. The information about the existing education system is as undernoted:

- There is dearth of good teachers
- Teacher–student ratio is not favorable
- Limitation of infrastructure and resources
- Pressure of fear and boredom in traditional classroom teaching-peer pressure
- Costly affairs to get a good education especially in technical and professional courses
- Most of the education system based on "one size fits all" producing only degree holders
- A large number of students across the world are needed to be educated
- There is a huge demand gap of skilled people in industry and education
- Time-consuming
- Running of obsolete academic and professional courses, etc.

These constraints are well catered and addressed by the e-learning courses. E-learning became very popular in a very short span of time, and it is indeed a ray of hope for future education system. It is very cost-effective, free from burden of maintaining huge infrastructure, source of modern-scientific-update learning materials, and epitome of anytime anywhere learning. It is inbuilt to exchange voice, text, pictures, and video seamlessly because of the availability of high-speed Internet connections in a cost-effective way. It fulfils the basic learning requirements by allowing learners to take center stage over the time and location choices. Moreover, adaptivity has been introduced in the e-learning courses. It is a proven fact that adaptivity has a significant effect on student's entire learning process. It has drastically changed the learning paradigm. It gauges the learners' needs and provides the materials as per their preferences and bridges gaps related to knowledge, skill, and attitude. It makes learning very interesting by providing learning objects in sequence of detected learning style preferences, and hence, learners themselves motivate to complete the course with a high note. It also enhances learning speed and maximizes learning outcome substantially. It also facilitates to improve overall educational management process and effectiveness of institutions in a cost-effective manner.

In this book, a framework for adaptive e-learning system has been developed and learning styles of the learners are identified by applying suitable educational data mining techniques based on Felder–Silverman Learning Style Model (FSLSM). EDM has been used because of its characteristics of the scalability, multidimensionality, and heterogeneity and capability in dealing complexity of the available data. In this model, automatic detection and adaptive e-learning system work together. Learners get the learning materials adaptively based on their activities and behavioural aspects captured in the log files as well as other databases. The captured data are associated with the learning objects and other collaborative activities in terms of the number of visits at a particular learning object, time spent at different learning objects, sequence of learning objects chosen by the learner, and percentage course completion in stipulated time. Here, characteristics of learning style play a crucial role in identifying them. They help in customization of learning material and delivery format in terms of responses made by the learners. The sharing and off-delivery of learning materials change dynamically and thus provide complete adaptivity.

Adaptive e-learning educational system individualizes the learning experiences for each learner. Learner's characteristics are quite different from one another's. Learners exhibit different behaviors and traits, form certain group of characteristics by which they are classified in different groups. Such classification helped system/tutor to provide the required learning materials as per choices made by the learners. It explores the way to achieve complete adaptivity which changes with respect to given variables. The framework is designed to understand how adaptivity works. The efficient adaptive framework has the capability to decide automatically what to deliver, how to deliver, and when to deliver. It is also referred to a model to mine learner's navigational accesses data and find learner's behavioral patterns. It individualizes each learner and provides personalization according to their learning styles in the learning process. Result and outcomes show that learners access relevant information effectively and efficiently according to their learning style. It is very useful in enhancing their learning process. This model is learner centric, but it also discovers patterns for decision-making process for academicians and people in decision-making bodies.

However, constructing an adaptive framework which identifies learner's learning style is the first small step toward the giant leap in terms of attaining our set objectives. Learning style refers to how learners learn and acquire knowledge in different ways. It enhances learning rate when input is provided in the way learner wants. The people may be classified differently according to their own style of learning. It can be understood better through learning style models. Some popular learning style models are studied; they are:

- Kolb's experimental learning model
- George's learning style model
- Felder–Silverman learning style model
- VARK's learning style model
- Dunn and Dunn's learning style model
- RASI's learning style model.

These models explain different measurable parameters of learning styles, analyze shared input information, and recommend suitable personalization. Among all the models, FSLSM is categorized as most popular, widely accepted as one of the best models, covers as many behavioral aspects of human being, and is considered as most appropriate for technical and engineering students. Therefore, it is chosen as a base model in our research study. It deals with the individual's characteristic strengths, weaknesses, and preferences in the way one takes in idea and process information (Felder & Silverman, 1988). It has a set of 44-item questionnaire which forms Index of Learning Style (ILS). It has five dimensions in which fifth one Inductive and Deductive has been dropped for engineering students, because it has proven that engineering students are inductive learners by nature. The other four dimensions are as undernoted:

- Active/Reflective: Active prefers doing things in group and doing work actively. Reflective prefers doing work alone and think before doing.
- Sensing/Intuitive: Sensing refers to facts and data obtained from experiments. Intuitive prefers ideas, theories, and innovations.
- Visual/Verbal: Visual prefers video, graph, picture, etc. Verbal prefers audio, text, etc.
- Sequential/Global: Sequential prefers step-by-step learning. Global prefers holistic approach learning.

Many adaptive e-learning systems do not take care of individuality of the learners and their distinct characteristics in a given situation. They do not bother about influences of learning style factors and readily present the same learning materials to all; e.g., one size fits all. Thus, the learning process experience and its effectiveness become less. To address these shortcomings and improve the efficiency of the learning process, the personalization in e-learning systems is developed to provide learning to each learner as per one's preferred learning styles. Adaptive system is the collection of adaptive techniques which enable system to adapt learners' learning style and suitably provide materials when required most.

The designed framework for adaptive e-learning system used EDM techniques. Intelligence system is introduced in the model by making suitable assumptions regarding LS preferences, knowledge levels, technical know-how, skills, attitudinal aspects, and learner's psychometric analysis. These tools are very helpful in determining what learner wants, and it also enables system or tutor to rise on the occasion to meet out the learning gaps dynamically and to achieve the set targets. There are many factors influencing the typical learning experience, and they translate into many more learning experiences "adaptive and personalized."

Our framework provides the concept of learner's model in the line of adaptive learning styles. They enhance the efficiency and effectiveness of the learning process significantly. Learners have completely different characteristics, and our model provides learning materials in tune with their own characteristics essential for learning in a way desired so. The characteristics consist of different knowledge levels, distinct cognitive abilities; individual's learning styles, levels of motivation, potent with self-efficacy factors, concentration levels, moods in the given situations and many more learning ability factors. All these parameters are taken into account, and adaptive materials are provided to learners. In addition, the complete adaptive process follows the dynamic path and continuously provides materials as per changes observed by the system automatically. The framework allows customization in consideration of individual's learning styles, capabilities to handle applications, processing materials, and reflecting learning information. Ultimately, the personalized response to each learner improves the learner's engagement and enhances the overall learning experience. Personalization in learner modeling has stipulated change in the area of adaptive e-learning applications, taking into consideration each learner's interests, preferences, and contextual information that vary widely.

The way human processes information is a matter of cognitive style. Cognitive style is thinking style in which individuals think, perceive process, and remember information in their own way. It is also related to individual's intelligence levels. Human brain does not analyze and segregate things in terms of crisp value. Rather it does in terms of related space, i.e., fuzzy values. It is understood that education-related studies, man–machine interaction in terms of input–output–decision making, and analysis of human behavior during learning are in the domain of soft computing. Therefore, soft computing techniques are used in this research work which is published in this book. In this research work, fuzzy cognitive map is most suited for analyzing adaptivity and recommending learning path. It is the combination of fuzzy logic and neural network. We use fuzzy cognitive map for knowledge representation and adaptive Neuro-Fuzzy Inference System (ANFIS) for implementation using Mental Modeler and MATLAB, respectively. Here, IF-THEN fuzzy rules in linguistic form depict adaptivity and allow suitable learning materials automated by considering all factors responsible for input variables.

The framework has provided personalization and tackles adaptation after applying soft computing techniques while acquiring, representing, storing, reasoning, and updating learner's profile and log data. The identification of learning style preferences and the recommendation of personalized adaptive model have been carried out based on fuzzy cognitive maps (FCMs) techniques. It is a fusion of soft computing techniques, i.e., a hybrid system combined with fuzzy logics and neural networks. FCMs deal in linguistic variables for automatically identifying learners' preferences based on their behavior and activities about the learning objects using LMSs in the adaptive e-learning environment. Adaptivity depends upon the selection of LOs, number of times accessed the object, time spent there and expected time, percentage of LO or course completed in the given time, and order of LOs or items selected. These measurable data are fuzzified in terms of linguistic variables similar to how human brain perceives in terms of very weak, weak moderate, strong, very strong, etc. The obtained data are again defuzzified into crisp values for understanding underlying facts. The implementation of these techniques has generated a large number of patterns which are mined to develop real-time assessment tool, decision support system, and adaptive delivery system to upgrade learner's academic performance and improve institute's overall effectiveness. The large amount of data involved and generated in the successful adaptation process creates complexity and possesses serious challenges. However, these data help to fine-tune the variables and help to search for further developments in the adaptivity process.

Learning is a continuous process since human evolution. In fact, it is related to life and innovations. The basic objective of learning to grow, aspire, and develop ease of life remains the same despite changes in the learning methodologies. Introduction of computers empowered us to attain new zenith in knowledge domain, developed pragmatic approach to solve life's problem, and helped us to decipher different hidden patterns of data to get new ideas. Of late, computers are predominantly used in education. Its process has been changed from offline to online in view of enhancing the ease of learning. With the advent of information technology, e-learning has taken center stage in educational domain. In e-learning context, developing adaptive e-learning system is buzzword among contemporary research scholars in the area of educational data mining (EDM). Enabling personalized is meant for improvement in learning experience for learners as per their choices made or auto-detected needs. It helps in enhancing their performance in terms of knowledge, skills, aptitudes, and preferences. It also enables speeding up the learning process qualitatively and qualitatively. These objectives are met only by the personalized adaptive e-learning systems in this regard.

Many noble frameworks were conceptualized, designed, and developed to infer learning style preferences, and accordingly, learning materials were delivered adaptively to the learners. Designing frameworks helps to measure learners' preferences minutely and provides adaptive learning materials to them in a way most appropriately. It is truly stated by Gerhard Fischer that "it is challenging to make available the bundle of information coming out across the world to the people at any time, at any place and in any form; but it is more challenging to say the right thing, at the right time and in the right perspective."

The benefits of e-learning over the traditional way of learning are well-known facts, and many world-class organizations have already implemented such courses mandatorily for their students or employees. It is very popular worldwide. It is cost-effective, consuming less time, promoting anytime–anywhere learning, supplementing non-availability of highly skilled teachers, and supporting learning by doing paradigm.

However, the e-learning courses offered by many organizations are outdated. Indeed, these courses are unable to cope up with the rising demand of personalized learning contents. Each individual learner is different from others and so the expectations of learning. The learning efficiency of the learners can be increased by providing personalized course materials and guiding them to attune with suitable learning paths based on their characteristics such as learning style, knowledge level, emotion, motivation, self-efficacy, and many more learning ability factors in e-learning system. In this regard, here we are explaining some terminologies which are very useful for readers to understand EDM and adaptive e-learning system.

This book discusses the main ideas introduced in and the stages of the research work in respective chapters. At first, an overview of the area of research work is discussed in which educational data mining, adaptive e-learning and its scope and challenges in education domain, and motivations are presented. Secondly, objectives of the research work and research issues related to this research work are given. In later part, sources of inspiration are described, followed by a discussion of the solution approach. Finally, the methodology and book contributions are described.

In broader aspects, our research area is related to educational data mining and adaptive e-learning system. So, it is an interdisciplinary area. This subsection gives a brief idea of area of research work. The book describes the escalated demands for e-learning courses and interlinked research work precisely. It represents numerous data mining techniques which are integrated with adaptive e-learning system to fully exploit the potential opportunities available in higher education system and to explore and re-discover the underlying knowledge from a large educational database.

Book contribution

- This book contributes to design a framework with its components containing different modules and to develop an adaptive e-learning system to meet out the criticalities of the existing adaptive e-learning systems and to counter its limitations.
- This book uses a FCM to model and construct the framework to implement adaptive e-learning system.
- Detection of learning style with learning ability factors generates learning style preferences through which personalized adaptivity is provided. Learning style detection makes learners know their learning style preferences and prepare them to use their strengths and weaknesses smartly.
- To develop a learner's model which incorporates learner preferences in adaptive LMS to provide personalization according to the learners' needs and to overcome the limitation of e-learning system.
- To develop an adaptive e-learning system, incorporating a noble framework and methodologies. The learner profile contains preferences and other relevant behavioral aspects that are responsible to provide personalized adaptive system.
- Learner preferences and requirements vary with time. Identify those variations and adapt them to a framework to get personalization and hence adaptation.
- To suggest educational management processes for improving overall effectiveness of institutions in a cost-effective manner.
- These adaptations include recommendations or feedback to students about their best next activities and changes as per their experiences and interactions with an online learning system.

Ranchi, India   Soni Sweta

Acknowledgments

First and foremost, I would like to express my sincere gratitude and profound regard to Chancellor **Dr. Atul Kumar Chauhan** and Vice Chancellor **Prof. (Dr.) Raman Kumar Jha** of Amity University, Jharkhand, for their continuous support during writing of this book and giving valuable suggestions and sharing knowledge. I am highly indebted to them and honestly thank them in this regard.

I express a very special thanks to Director Prof. (Dr.) Ajit Kumar Pandey for his incredible suggestions and constant encouragement during completion of this book. I owe my eternal gratitude to Prof. (Dr.) D. K. Singh, Director, BIT Sindari; Dr. M. P. Singh, NIT Patna; Prof. (Dr.) Sanjeev Sharma, Dean, RGPV, Bhopal; Prof. (Dr.) S. P. Lal, BIT Mesra, Patna; and Dr. Kanhaiya Lal, Associate Professor, BIT Mesra, Patna. It was under their tutelage that I found right direction, distinct approach, and perfect blend to complete this book.

My sincere thanks also go to Amity University, Jharkhand, and its entire team that encourage me to write a book, access to the library, laboratory, online resources, research facilities, enduring technical support and all facilities required to complete epitome of this book. Without its precious support, it would not be possible to write this book.

I express deep sense of my gratitude to all faculty members of Amity University, Jharkhand, especially Dr. Joyeeta Chatterjee, Dr. B. Samanta, Dr. Pooja Jha, library in-charge, and all other peoples who directly and indirectly supported me during completion of this and for their stimulating motivation, enlightening moments, moral support, and constant encouragement.

Last but not least, I would like to thank my family members: especially my mother **Prof. (Dr.) Sheo Kumari Singh**, my elder sister **Dr. Sangeeta**, my brother-in-law **Dr. Rajiv Ranjan**, my elder brother **Mr. Kumar Abhishek**, my husband **Mr. Daya Shankar**, my younger brother **Mr. Kumar Saurabh**, and my son Master **Arunabh Kshitij** for supporting me spiritually throughout writing this book, sacrificing their life, enduring patience and my life in general, without whose love and encouragement, I would not have finished this book.

Thanks to almighty God to give me strength and confidence to complete this book on time.

Soni Sweta

Contents

1 Educational Data Mining in E-Learning System 1
   1.1   Introduction to Educational Data Mining (EDM) 1
   1.2   Technology Enhancement Educational Data Mining 1
   1.3   Applications of EDM 3
   1.4   Data Mining Outlook in Business and Educational Domain 4
   1.5   Terminology Used in EDM 4
   1.6   E-Learning System 6
   1.7   Evolution of Education System 6
   1.8   Research Area in E-Learning 7
   1.9   Limitation of E-Learning 10
   References 11

2 Adaptive E-Learning System 13
   2.1   Overview of Adaptive E-Learning System 13
   2.2   Adaptive E-Learning 13
   2.3   Adaptive E-Learning Systems 14
   2.4   Intelligent Versus Adaptive 14
   2.5   Adaptive E-Learning in Terms of Data Mining 16
   2.6   Application Area of Adaptive E-Learning 16
   2.7   Adaptive Parameters in E-Learning? 16
   2.8   Adaptive LMS System Functionality (SF) 17
   2.9   Learning Management System 18
   2.10   Adaptivity in Learning Management Systems 18
   2.11   Kinds of Adaptivity 18
   2.12   Why Adaptation is Required? 18
   2.13   Personalization with Adaptivity 20
   2.14   Adaptive E-Learning Its Scope and Challenges 20
   2.15   Scope—A Tool to Manage Shortcomings of E-Learning 21
   2.16   Some Existing Challenges 21
   2.17   Process of Adaptation 22
   References 22

3 Educational Data Mining Techniques with Modern Approach 25
  3.1   Introduction 25
  3.2   Data Mining in Terms of Adaptive E-Learning and Web
        Personalization 25
        3.2.1   Soft Computing Techniques in Data Mining 26
        3.2.2   Advantages of Soft Computing 26
  3.3   Soft Computing Techniques in Personalized Adaptive
        E-Learning System 27
        3.3.1   Comparative Analysis of Expert Systems, Fuzzy
                Systems, Neural Networks, and Genetic Algorithms 28
        3.3.2   Intelligent Hybrid System 28
        3.3.3   Neuro-Fuzzy Approaches 29
        3.3.4   Advantages of Combination of Neuro-Fuzzy 29
  3.4   Fuzzy Cognitive Map (FCM) 29
        3.4.1   Application of Fuzzy Logic in Education 30
        3.4.2   Advantages and Disadvantages of Fuzzy Logic 30
  3.5   Data-Driven Approach Versus Literature-Based Approach 31
        3.5.1   The Data-Driven Approach 31
        3.5.2   The Literature-Based Approach 32
  3.6   Learner Modeling Techniques for Personalized and Adaptive
        E-Learning 33
        3.6.1   Bayesian Belief Network 33
        3.6.2   Fuzzy Logic-Based Technique 34
        3.6.3   Neural Network-Based Techniques 34
        3.6.4   Fuzzy Clustering-Based Techniques 35
  3.7   Learning Style-Based Individualized Adaptive E-Learning 35
  References 36

4 Learning Style with Cognitive Approach 39
  4.1   Learning Style with Learning Theory 39
        4.1.1   Learning Style 39
        4.1.2   Learning Styles Theories 39
        4.1.3   Six Prominent Learning Style Models 40
        4.1.4   Discussion on Existing Personalized Adaptive
                E-Learning System 40
  4.2   Comparative Analysis of Learner Modeling Techniques 47
  References 48

5 Framework for Adaptive E-Learning System 51
  5.1   Introduction 51
  5.2   An Adaptive Framework 52
        5.2.1   Learner Profile and Interface Module (LPIM) 53
        5.2.2   Behavior Monitoring Module-BMM 56
        5.2.3   Learning Style Diagnostic Module-LSDM 58
        5.2.4   Personalized Adaptive Module-PAM 59

5.3 Workflow of Adaptive System Components 60
5.4 Adaptable Characteristics for Learner Model with System
    Process 61
5.5 Factors Affecting on Personalization 61
5.6 Factors Affecting Adaptation 62
References 62

6 Personalization Based on Learning Preference 63
6.1 Introduction 63
6.2 Fuzzy Cognitive Maps 63
    6.2.1 FCM Outlines and Narration 64
    6.2.2 Knowledge Representation with FCM 64
    6.2.3 Different Mathematical Representation of Fuzzy
          Cognitive Maps 65
    6.2.4 Structure and Learning from FCM Method I 66
    6.2.5 FCM: Method II 66
    6.2.6 FCM: Method III 67
6.3 The Felder–Silverman Learning Style Model 68
    6.3.1 Scales of the Dimensions 68
    6.3.2 Learner Preferences in Five Dimensions 69
    6.3.3 Index of Learning Style (ILS) Scale 69
    6.3.4 Combination of Learning Style 70
    6.3.5 Annotating Learning Objects 71
6.4 Learner Modeling Based on Fuzzy Cognitive Map (FCM) 72
    6.4.1 Diagnose Learner Profile 72
    6.4.2 Collecting and Processing Information 73
    6.4.3 Input Layer 74
    6.4.4 Output Layer 74
6.5 Labeling Learning Objects 75
6.6 Automatic Detection of Learner's Characteristics in E-Learning ... 75
    6.6.1 Behavior Monitoring Module-BMM 75
    6.6.2 Auto-Diagnosis of Learner's LS in E-Learning 77
6.7 Measurements of Concepts Values and Their Strengthens
    Casual Relationship 82
6.8 Fuzzy Rule Base in Terms of Adaptive Rules 84
References 85

**7 Recommender System to Enhancing Efficacy of E-Learning
System** 87
7.1 Introduction 87
7.2 Personalized Adaptive Module-PAM 88
    7.2.1 Calculating Motivation (High/Low) 88
    7.2.2 Calculating Knowledge Ability (Low/Medium/High) 88
7.3 Overview of Implementation of a Novel Framework (NAPF) 90

7.4   Fuzzy Inference Systems (FIS) 90
       7.4.1   Sugeno Fuzzy Inference Method 91
7.5   ANFIS Editor and Training 92
7.6   Model Training 92
7.7   Validation of Model 93
References 93

About the Author

Dr. Soni Sweta received her Master of Technology from Rajiv Gandhi Proudyogiki Vishwavidyalaya, Bhopal, in 2011. She has successfully completed her Ph.D. in the area of soft computing and data mining in computer science and engineering from Birla Institute of Technology Mesra, Ranchi, in 2018. She has published more than 12 research papers in high-impact peer-reviewed international journals, 2 papers in national journals, and 4 book chapters. Her research areas are soft computing, artificial intelligence, machine learning, data science, and data mining. From 2004 to 2010, she has worked as Assistant Professor and Visiting Faculty in computer science and engineering departments of different engineering colleges, including MLB College; SNGPG College; Scope Engineering College, Bhopal; NSIT Patna; and CIPET Lucknow. She is presently working as Assistant Professor in Amity University, Jharkhand, Ranchi, India.

Abbreviations

Act	Active
ADT	Abstract data type
AEH	Adaptive education hypermedia
AIWBES	Adaptive and intelligent web-based educational system
ALS	Adaptive learning systems
ANFIS	Adaptive Neuro-Fuzzy Inference System
ANN	Artificial neural network
APIE	Adaptive personalized intelligent e-learning
APIES	Adaptive and personalized intelligent e-learning system
AWBES	Adaptive web-based education system
BBN	Bayesian belief network
BMM	Behaviour Monitoring Module
BN	Bayesian network
BS Test	Behavioral science and soft skill development
CI	Computational intelligence
CMC	Combination of multiple classifiers
CS	Cognitive style
Dim	Dimension
DLA	Distance learning
DM	Data mining
DT	Decision tree
EC	Evolutionary computation
EDM	Educational data mining
eLMS	Electronic learning management system
ES	Expert system
FCM	Fuzzy cognitive map
FD	Field dependent
FI	Field independent
FIS	Fuzzy inference system
FL	Fuzzy logic
FS	Fuzzy system
FSLSM	Felder–Silverman learning style model

GA	Genetic algorithm
Glo	Global
HCI	Human–computer interface
HIS	Hybrid intelligent system
HMM	Hidden Markov model
ICT	Information and communication technology
ILS	Index of Learning Style
Int	Institutive
IT	Information technology
ITS	Intelligent tutoring system
KAF	Knowledge ability factor
KDD	Knowledge data discovery
KM	Knowledge management
KNA	Knowledge ability
LAF	Learning ability factor
LB	Learning behavior
LCMS	Learning content management system
LMS	Learning management system
LO	Learning objects
LP	Learning preference
LPIM	Learner Profile and Interface Module
LS	Learning style
LSDM	Learning Style Diagnostic Module
MLE	Managed learning environment
MLP	Multilayer perception
MOODLE	Modular object-oriented dynamic learning environment
MT	Motivation
NAPF	Novel adaptive personalized framework
NN	Neural network
Oop	Object-oriented programming
PALM	Personalized Adaptive Learner Model
PAM	Personalized Adaptive Module
Ref	Reflective
SCORM	Sharable content object reference model
Sen	Sensing
Seq	Sequential
SF	System functionality
Sim	Similarity
Ver	Verbal
Vis	Visual
VLE	Virtual learning environment

Symbols

$\leftarrow$	Implication
$\sum$	Summation
μ/m	Membership function
f	Function
C_i	Concept
w_{ij}	Weight on edge from C_i to C_j
Pc_i	Concept value on C_i
k	Constant
%	Percentage
$\in$	Member
Θ	Theta
∞	Infinity
ε	Epsilon
$\leq$	Less than or equal to
$\geq$	Greater than or equal to
$\neq$	Not equal to
X	Input value
O	Output value
k1	Concept dependence
k2	New concept value parameter
f	Sigmoid function
n	No. of term
S	Set of students
T	Learning activity tool
B	Set of behaviour patterns
A	Fuzzy linguistic term

Chapter 1
Educational Data Mining in E-Learning System

1.1 Introduction to Educational Data Mining (EDM)

Educational data mining (EDM) is the most emerging academic and research field of computer science which extracts, explores, and exploits data to infer valuable information in the educational context by using data mining (DM) techniques [1]. It incorporates the application of data mining, machine learning, and statistics using artificial intelligence tools.

It bundles out many new ideas in education like how to measure learners' needs, how to teach them well, what learning materials provided to them, and what learning processes chosen for them for effective educational planning and implementation [2]. Table 1.1 shows the different types of stakeholders with their specifies purposes:

How these stakeholders work together in an education system and where educational data mining techniques implemented for pattern analysis are also shown in Fig. 1.1.

1.2 Technology Enhancement Educational Data Mining

As the entire education system has been transforming from print to digital and correspondingly learning products has also been changed rapidly, the academic institutions and policies have responded to the inevitable changes accordingly [3]. Learning methods are continuously shifting from teacher centric to learner centric which refers to the increasing trend of learning via technological tools like multimedia/Internet. Learner-oriented learning provides to the point delivery, smart way of learning, improves learning performance, and enhances satisfaction matrix. It facilitates learning-on-fly mechanism and of course, instrumental in leveraging the technology in view of learning time and cost. Instructor or system prepares

Table 1.1 EDM with different types of stakeholders

Users	Objectives for using educational data mining
Learners/Students	Personalized e-Learning and recommendation
Educators/Teachers/Instructors/Tutors	Feedbacks from learners and management
Course Developers/Educational Researchers	Evaluation and maintenance of courseware
Organizations/Learning Providers/Universities/Private Training Companies	Enhancing decision making in higher learning institutions
Administrators/Network or System Administrators	Carving the system optimally to use and organize available institutional resources and their educational offers

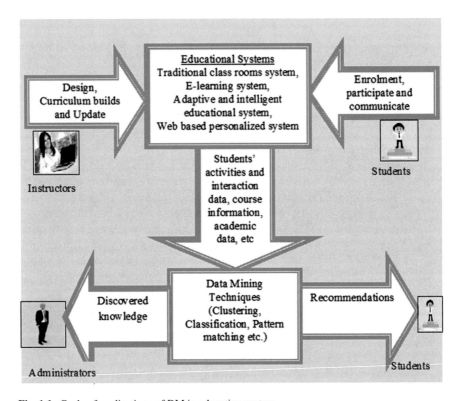

Fig. 1.1 Cycle of applications of DM in education system

learning materials and customizes its packaging dynamically keeping in view of fly-on-demand.

With the ever-increasing utility of computers and new researches in telecom revolution, the Internet/Intranet as medium for networking has supported transmission of skills, information, and knowledge beyond boundaries [4]. E-learning is Internet-based education system in which e-teaching is provided either in form of asynchronous, synchronous, and hybrid way. It is a new context for e-education system where a large amount of data and information is readily made available online. It is the continuum of the teaching–learning interactions in the prescribed formats to scale up the new heights. It is in fact a blessing in disguise because voluminous information is readily available just at fingertips. However, it is equally a nightmare if it is not channelized properly. The unstructured and uncontrolled information cause deviation in the educational system. They impede transmission of knowledge and hinder the interest of all stakeholders.

Some e-learning problems have been suitably addressed by data mining techniques [5] which are discussed as follows: students' classification based on their profile data, inherit qualities, learning preferences, learning performance, etc. Some more advanced findings consist of detection of irregular learning behaviors; optimization of e-learning system in terms of navigation and interaction; clustering based on similar e-learning system usage; and systems' adaptability to learner's needs and capacities. The assessment of learners' e-learning issues is easily handled by way of data mining techniques. Such assessment is much closer to the evaluation mechanism available in the traditional education system.

The popularity of the internet has enabled online distance education to become reality in mainstream education system. It happened only because of many new e-learning models and systems have been developed, and the core findings have been implemented with varying degrees of success. Obviously, the inclination of learners toward the new technologies has reinforced the efforts made by e-learning researchers and developers in EDM. They produce phenomenal strings of data and information which help to improve the e-learning system by using deciphered knowledge patterns into valuable information. Data mining concepts are used to extract such knowledge patterns. Its main objective is deriving the information from the domain of patterns of system usage in which the most important ones are finding the learner's learning behavioral patterns and analysis.

1.3 Applications of EDM [6, 7]

- Analyzing and recapitalizing data patterns
- Arranging suggestions and feedbacks for better orientation of instructors or systems
- Informing learners to improve their approaches suitably
- Predicting student performance or success or promotions
- Creating models for individual learner or group of learners
- Detecting undesirable learner's behaviors from internal system as well as from various available social platforms
- Introducing cognitive maps or designing courseware

Table 1.2 Comparison between business and educational domain

Business domain	Educational domain
Quest for most profitable customers	Quest for learners consuming most credit hours
Quest for the most frequent customers	Quest for ones who return for more classes
Quest for loyal customers	Quest for consistent learner
Quest for customers likely to purchase	Quest for alumnus likely to donate/pledge more
Quest for the customers likely to defect to my rivals	Quest for courses that attract more learners

- Structuring and scheduling in different educational system like traditional, web-based, intelligent tutoring, and adaptive e-learning system
- Formulating recommender system based on EDM helps learners to take better decision for choosing best courses/facilities from the pool of available ones [8].

1.4 Data Mining Outlook in Business and Educational Domain

Generally, data mining techniques used in business domain to taking better decision, but it has more popular in education sector also as a business intelligence. Table 1.2 shows the types of user and their needs of information in business and education domain.

1.5 Terminology Used in EDM

Some terms and terminology are used in this book for better understanding. The most frequently used words are as follows:

- **Learning Gaps**: A gap is a discrepancy or anomaly between what is there and what should be [9]. So, **learning gap** is the difference between what we should know about how to maximize effective learning and what is currently being providing or happening in the classroom, i.e., the way the brain learns itself and the way learners are taught in the classroom.
- **Types of Learning Gaps**:
    i. **Individual**: Personal development program to develop required level of knowledge, skills, aptitudes, and preferences
    ii. **Operational**: Fulfilling institute's objectives, shared vision, team cohesiveness, common set of core values, etc.

iii. **Organizational**: Strategic planning, review, control of performance indicators etc.

i. **Individual Learning Gaps**: Initially, learners are in indecisiveness which course is the best suitable alternatives for them and which one to be chosen finally. It is very difficult jinx to guide without a proper recommender system. Even excellent teacher does not guarantee student's success across any educational systems. It depends upon student's individualities, traits, and characteristics. Individual gaps may be addressed suitably:

- **Knowledge Gaps**: They are detected by different evaluation methods like exam test, quiz, and assignments. In digital media, computers and consoles are cognitively very powerful instruments to detect the gap.
- **Skills Gaps**: They are detected by competencies, i.e., workplace utility increased by exercise, lab simulation, case study, applications, etc.
- **Aptitude and Preferences Gaps**: Quantitative and observable evidences are measured through the trainings/sessions related to behavioral science and soft skill development (BS Test).

- **Learning Style Theory**:
  Learning theories make congruence to the fact that learners learn and attain knowledge in many ways. Such differentiation comes from classification in learning styles. One of LSs' classifications have been proposed by Felder and Silverman [10], Kolb [11] and many other scholars [10].
- **Learning Process**:
  The distinct nature of learners having particular learning style and mode for thinking and different abilities in relating and creating things are factors responsible for individuality [12]. It has implications over the teaching strategies adopted by the instructor/system. Learning process improves significantly in tuned teaching into its leaner's preferred modes because the preferred modes of inputs vary from individual to individual. As output depends upon input variables, they also change accordingly.
- **Learning Behaviour**:
  It is related to the ever-changing psychological–emotional state of the mind while interacting with the e-learning system. Sometimes, it is deliberately carried out and sometimes, it is based on stimuli. Boredom, frustration, motivation, concentration, tiredness, etc. are emotional quotients which are identified as important behavioural aspects enabled with to make or mark the effectiveness of the learning process [13].

Identifying and tracking learning behaviors in real-time scenario is a very important job. Conati and Gutl et al. in [14] proposed theories to modeling learner's influence and structured the concepts to evaluate various emotional phases during the interactions. They applied special sensing scanners to detect behavior (following the Ortony, Clore and Collins cognitive theory of emotions). Milosevic et al. [15]

observed that the learners' motivation plays a crucial role in ensuring learning efficiency in many learning theories. It is interlinked to emotionally charged, self-inclined, focused, high level of concentration, worried, interactively by giving feedback, system generated recommendation, etc. A technique used in "Discovering student preferences in e-learning" by Carmona et al. [16] is interesting study to implement to find out learner's preferences.

1.6 E-Learning System

E-learning is an alternative system in which learning process takes place predominately in the electronic format. E-learning occupies the nomenclature in terms of VLE—Virtual Learning Environment, e-LMS—Electronic Learning Management System, LCMS—Learning Content Management System, MLE –Managed Learning Environment [17]. VLE is a computer software system in which word, video, audio, animation, network, etc., are used in the learning process. It can be asynchronous or synchronous.

E-learning system is multidisciplinary by its nature. Many researchers studying computer science and application, information technology, psychology, education system, and educational technology have been trying to evaluate and find new things about e-learning systems in their respective domains. Some of them had emphasized on technology-driven components of e-learning systems, whereas some had focused mainly the human factor in e-learning systems. Arbagh and Fich [18] stressed upon the importance of man–machine interactions in online systems, and Gilbert [19] investigated the learner's experiences in detail. E-learning is defined in terms of the learning process via electronic media only. According to Rosenberg [20], e-learning depends on IT and related environment. It is typically a networked learning system which overcomes the traditional way of learning. E-learning consists of all forms of electronically supported learning materials and system-generated teaching based on system-generated log files. It works better in any networked environment based on Intranet or Internet.

1.7 Evolution of Education System

- **Traditional Classroom System**: Teacher teaches students in a classroom. It is an example of face-to-face learning and in interactive way. The process takes place at the same time and the same place. Learners get the information all about what teacher teaches in a classroom and how teacher is prepared well before taking the session. Performance evaluation examinations are conducted at different locations for assessment and certificate issued by the authorities. It is a teacher-centric approach, and learning process is limited to content delivering in the classroom.

- **Distance Learning System**: Students do correspondence courses in which students are located at remote places, get study materials from institute at their own locations, do assignments at their locations, and send them to the institute for assessment. Students participate in final exam for getting certificate at any specified center like the traditional system.
- **E-learning System**: With the help of Internet/Intranet, learners do online courses after registration in which learners are located across the world [21]. Here, teacher and learner need not required to be present at same location and at same time. They complete their courses online; ask queries; and clear doubts through chat/forum/e-mail. They learn the courses mainly with the help of multimedia tools, do simulations in course materials, if made available and participate in assessment test online or offline as per specified policy of the institute at term end. It is very popular way of learning. It is in great demand in the educational domain because here state-of-the-art technology plays role as most effective enabler. It is also predominantly teacher-centric approach, and hence, this system is unable to meet the expectations on the scale of learner's satisfaction.
- **Intelligent Web-Based System**: It is an improved version of e-learning courses in which tutor provides learning materials on demand put through by the learners via online system in addition to the course materials already made available at the portal. In this system, a single tutor manages, monitors, and moderates a large number of learners located at far-away distant locations [22, 23]. The learners like such interactions, involve in discussion, and show their interest to get more and more information from the tutor. Online or offline assessment tests are conducted as per policy of the institute. Figure 1.2 shows the step-by-step evolution of education system.

1.8 Research Area in E-Learning

E-learning is recently emerged as a very advanced research area. Here, a few prominent and interesting research areas are discussed as under:

- **E-learning Effectiveness**:
  E-learning has been grown like a wildfire with support of the new economic scenario, and it has immense potential to impart effective learning using state-of-the-art technologies. Many of the researches in this area are based on web-related learning, and they described how its effectiveness can be maximized. They also compared studies with the traditional classroom learning, i.e., brick-and-mortar system and other popular learning systems. The researchers had assessed effectiveness of e-learning by analyzing post course questionnaire submitted by learners, observing data log file patterns of online activities, gathered valuable information from verbal and written deliberations, and compared their performance based on tests, course grades, and overall performance during the entire learning process.

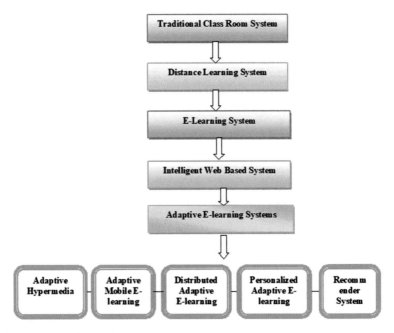

Fig. 1.2 Evolution of education system

- **E-learning Based on Multimedia Techniques**:
  It has been already proven in various researches that the multimedia techniques enhance individual's performance in terms of problem solving, utilizing skills, and seeking full attention during content delivery. Because of the vividness of the presentation and usage of audio-visual tools, create mesmerizing environment which invites the active participation and great fascination; truly, multimedia helped immensely.
- **Technological Support to E-Learning**:
  Multimedia technologies, electronic distribution on web pages, and real-time online adaptive methods have paved the new ways in development of e-learning. It had grown rapidly during the last decade.
- **Internet/Web-based E-Learning (Synchronous/Asynchronous)** [18]:
  As of now, the unprecedented changes in technology have created wide scope around education and training. There is a huge opportunity for the learner's vis-a-vis the organizations for providing e-learning portals. Many new green field activities are being done in this field to meet out the rising expectations. It is really challenging to fulfill outbound expectations.
- **Dynamic E-Learning**:
  Of late, the delivery of multimedia course contents as demanded by the user working with the personnel computer is simultaneously supported by the rampant

use of Internet. It has created a big platform for adaptive and interactive video technology to enhance the perception between the systems, i.e., the virtual instructors and the learners. The audio-visual effects attract more learners' engagement by its spellbound charms, and hence, it enhances their performance substantially [24]. It creates less hindrance in execution of task. We can easily compress a video file and play it while it is being downloaded to reduce video transmission delay. It also enhances interactive e-learning that is one-to-one interaction as good as classroom learning. Retrieval mechanism and easy-to-use interface have ensured maximum utilization of bandwidth and efficient e-learning effective video indexing.

- **Business Intelligence**:
  It is a knowledge management (KM) in which optimization, control, and management are the most critical production factors. Business intelligence helps employees to become professionally efficient and eventually effective for their own organization. In another way, knowledge management is all about collecting, managing, and sharing knowledge, which is a prerequisite requirement for gaining knowledge, commanding overseas acquisitions, and again sharing enhanced knowledge. The biggest knowledge management challenge in e-learning courses remains the same to meet out demand of a learner's needs. It can only be provided through efficient management accessible, highly distributed information, and knowledge system through dynamic adaptivity [25].

- **E-commerce Technologies**:
  The superimposition of technology and commerce has created a large potential for course providers to earn handsome revenues. It is in fact indispensable collaboration that is needed to be promoted in view of increasing cost of education and the learners' indifferent approach toward the existing learning system. Its time and cost-effective nature encourage identifying the shortcomings and updating the ongoing e-learning systems for covering more courses and enable to reach the learners across the world.

- **Collaborative E-Learning**:
  A virtual classroom is a platform in which learners and instructors though remotely placed, they create a learning platform of collaborative learning and take leverage of the latest technology. Many research studies have proven that collaborative learning induces better learners' participation, better group synergy, better academic pursuits, and improve overall productivity in comparison to the individual learning [18]. Some of the widely used collaborative tools are—Electronic Bulletin Boards, News Groups, Net Meeting, Online Chat rooms, and Web-Based Group System. Text-based online collaboration tools are mostly used for supporting e-learning systems. But, tools based on multimedia are highly demanded in the existing system of adaptive e-learning [18].

- **Digital and Telecommunication Technology**:
  There is a significant change on the evolution of e-learning in digital world. Now, instructors and learners are easily accessing the reliable digital e-resources, and they can build high-quality distributed learning communities [18]. The high-speed networking and high-resolution videos introduced into e-learning system have

improvised the system further. Telecommunication enabled learners to communicate with peers and instructors by using databases and other information sources effectively and seamlessly.

- **Human–Computer Interface**:
  It is true that different users located in different part of the world have different conceptions and perceptions about interactions with the computer. It may be because of variations in terms of culture, education, and accessability to computer and digital world. The interface preferences change over a period. The most studied five aspects of human–computer interface are comprising the nature of HCI, human characteristics, the user and context of computers, computer system and interface architecture, and development processes [17]. The learners participate in online discussions are very often provided with the easy access of Internet.
- **Evaluation of E-Learning**:
  It is necessary to assess learner's progress during a learning process. Moreover, it is equally important to evaluate the system's performance before evaluating the existing e-learning system.
- **Security and Accounting Aspect of E-Learning**:
  E-learning system must be ensured against the malicious access. It may lead to misuse of the data. It must protect the privacy of learners and instructors. It must have adequate support system to allow access only after due verification of the authorized users only. It must ensure the obligations in terms of worldwide copyright and license agreements about electronic learning materials [3]. It must be made in such a way to avoid illegal uses. Although the encryption technique and digital signature method are used to protect data, its misuses cannot be ruled out.

1.9 Limitation of E-Learning

It is a proven fact that text-based learning material has not enough content for ensuring better understanding in comparison to materials based on multimedia. But unstructured and isolated multimedia content also does not motivate many researchers as well as learners. On examining the impacts of learner's characteristics on e-learning effectiveness, powerful simulation and experimental opportunities are embedded to enable learners to acquire new knowledge, skills, etc. There is a lot of research on intelligent tutoring and adaptive e-learning system that can be applied in this area. Concerted study is required how to efficiently assess learners' performance and make dynamic adjustment in the instructional contents provided to the learner. It is necessary to investigate the impacts of different learning contexts on e-learning effectiveness. In other words, it is also important to identify what type of content is more suitable for online learning and what can be ignored deliberately.

References

1. Peña-Ayala A (2014) Educational data mining: a survey and a data mining-based analysis of recent works. Expert Syst Appl 41(4) PART 1:1432–1462. https://doi.org/10.1016/j.eswa. 2013.08.042
2. Baker RSJD, Yacef K (2009) The state of educaitonal data mining in 2009: A review and future visions. J Educ Data Min 1(1):3–17, 2009. https://www.educationaldatamining.org/JEDM/ima ges/articles/vol1/issue1/JEDMVol1Issue1_BakerYacef.pdf
3. Sweta S, Lal K (2014) Adaptive e-Learning system: a state of art. Int J Comput Appl 107(7):13–15. https://www.ijcaonline.org/archives/volume107/number7/18762-0046
4. Romero C, Ventura S (2010) Educational data mining: a review of the state of the art. IEEE Trans Syst Man Cybern Part C (Applications Rev) 40(6):601–618. https://doi.org/10.1109/ TSMCC.2010.2053532
5. Jindal R, Borah MD (2013) A survey on educational data mining and research trends. Int J Database Manag Syst 5(3):53–73. https://doi.org/10.5121/ijdms.2013.5304
6. Romero C, Ventura S (2013) Data mining in education . Wiley Interdiscip Rev Data Min Knowl Discov 3(1):12–27. https://doi.org/10.1002/widm.1075
7. Özyurt Ö, Özyurt H (2015) Learning style based individualized adaptive e-learning environments: content analysis of the articles published from 2005 to 2014. Comput Human Behav 52:349–358. https://doi.org/10.1016/j.chb.2015.06.020
8. Zapata A, Menéndez VH, Prieto ME, Romero C (2015) Evaluation and selection of group recommendation strategies for collaborative searching of learning objects. Int J Hum Comput Stud 76:22–39. https://doi.org/10.1016/j.ijhcs.2014.12.002
9. Lucke U, Rensing C (2013) A survey on pervasive education. Pervasive Mob Comput 14:3–16. https://doi.org/10.1016/j.pmcj.2013.12.001
10. Felder R, Silverman L (1988) Learning and teaching styles in engineering education. Eng Educ 78(June):674–681. https://doi.org/10.1109/FIE.2008.4720326
11. Kolb DA (1981) Learning styles and disciplinary differences. In: Responding to the new realities of diverse students and a changing society, pp 232–255. https://doi.org/10.1016/S0002-822 3(97)00469-0
12. Sweta S (2015) Adaptive and personalized intelligent learning interface (APIE-LMS) in e-learning system 10(21):42488–42492
13. Truong HM (2015) Integrating learning styles and adaptive e-learning system: current developments, problems and opportunities. Comput Human Behav. https://doi.org/10.1016/j.chb. 2015.02.014
14. Botsios S, Georgiou D (2008) Recent adaptive E-Learning contributions towards a "standard ready" architecture. e-Learning. https://utopia.duth.gr/~dgeorg/PUBLICATIONS/53.pdf
15. Milošević I, Živković D et al (2015) The effects of the intended behavior of students in the use of M-learning. Elsevier, Computers in Human, DM-C in H. https://www.sciencedirect.com/sci ence/article/pii/S0747563215003362
16. Carmona C, Castillo G, Millán E (2007) Discovering student preferences in e-learning. CEUR Workshop Proc 305:33–42. 10.1.1.30.9978
17. Dominic M, Xavier BA, Francis S (2015) A framework to formulate adaptivity for adaptive e-learning system using user response theory. I J Mod Educ Comput Sci 1:23–30. https://doi. org/10.5815/ijmecs.2015.01.04
18. Arbaugh JB, Benbunan-Fich R (2007) The importance of participant interaction in online environments. Decis Support Syst 43(3):853–865
19. Gilbert J, Morton S, Rowley J (2007) e-Learning: the student experience. Br J Educ Technol 38(4):560–573. https://doi.org/10.1111/j.1467-8535.2007.00723.x
20. Sampson D, Karagiannidi C, Kinshuk (2002) Personalised learning: educational, technological and standardisation perspective. Interact Educ Multimed 4(4):24–39. https://greav.ub.edu/iem/ index.php?journal=iem&page=article&op=view&path[]=26&path[]=24
21. Urdan TA, Weggen CC (2000) Corporate E-learning : exploring a new frontier. Analysis, p 88. https://wrhambrecht.com/research/coverage/elearning/ir/ir_explore.pdf

22 Mohammad Bagheri M (2015) Intelligent and adaptive tutoring systems: how to integrate learners. Int J Educ 7(2):1. https://doi.org/10.5296/ije.v7i2.7079

23. Stathacopoulou R, Magoulas GD, Grigoriadou M, Samarakou M (2005) Neuro-fuzzy knowledge processing in intelligent learning environments for improved student diagnosis. Inf Sci (Ny) 170(2–4):273–307. https://doi.org/10.1016/j.ins.2004.02.026

24. Gonzalez-Brenes JP, Mostow J (2012) Dynamic cognitive tracing : towards unified discovery of student and cognitive models Jos e. In: Proceedings of 5th international conference on education data mining, pp 49–56

25. Chen H, Chiang RHL, Storey VC (2012) Business intelligence and analytics: from big data to big impact. Mis Q 36(4):1165–1188

Chapter 2
Adaptive E-Learning System

2.1 Overview of Adaptive E-Learning System

Nowadays, many e-learning frameworks are made available to provide different types of e-learning architecture and realizing different experiences. But most of them are related to the concept of "one size fits all" [1]. To overcome such limitations in all educational system from traditional to web-based intelligent tutoring system, adaptive and personalized e-learning system is a ray of hope. Adaptive e-learning is an educational method in which Internet-enabled computers are used as interactive teaching devices loaded with the required software functionalities [2] which create virtual classroom sort of teaching between learner and teacher. In this system, computers primarily adapt the presentation of educational materials according to learner's learning needs automatically detected by the system and recorded in log files on basis of learner's activities and interactions with the learning objects as given in the course. Here, computer is a tool that plays the important role of enabler to provide adaptive environment (Fig. 2.1).

Adaptive hypermedia is a relatively recent in research area related to hypermedia and user modeling. Adaptive hypermedia and its significance in stage-wise updating of e-learning system are exhaustive, and it can be studied separately. The work on the third-generation adaptive educational hypermedia eventually leads to complete integration of adaptive technologies into the process of web-based learning which maximizes the ability of learner to achieve the targets [3].

2.2 Adaptive E-Learning

Adaptive e-learning provides a way of adaptivity in which learners have the choices to fix the time, location, preferences, and demand at their own will to get the desired learning materials at the time of learning itself. Therefore, the e-learning system

© The Author(s), under exclusive license to Springer Nature Singapore Pte Ltd. 2021
S. Sweta, *Modern Approach to Educational Data Mining and Its Applications*,
SpringerBriefs in Computational Intelligence,
https://doi.org/10.1007/978-981-33-4681-9_2

Fig. 2.1 Evolution of adaptive E-learning system

begins to evolve around adaptable e-learning system where learners experience customized and personalized learning process. It infers the learners' characteristics to identify the learner's preference to generate personalized learning path and customized learning materials to the learners. After that it will be also benefited for making a recommender system in adaptive e-learning environment.

2.3 Adaptive E-Learning Systems

It is an improvised and more advanced e-learning system in which learning style and other behavioral aspects like motivation, emotions, etc., are auto-detected and factored in providing learning materials as per choices made by the learners knowingly and unknowingly or automatically provided by the system on the basis of auto-detected log files [4]. Adaptivity enhances learning speed, learning process, learning experience, learning outcome and improves overall learning effectiveness [5].

Adaptive e-learning system is the world of collection of all techniques which enable and lead to adaptation in a web-based e-learning. It can be better understood by Fig. 2.2 which shows the origin of Adaptive Web-Based Education System.

2.4 Intelligent Versus Adaptive

An intelligent tutoring system (ITS) is a computer-added system which provides customized instructions and recommendations to learners immediately [6] mostly without human intervention. ITSs are aimed to enable learning in a meaningful and effective way by using computing technologies.

The traditional intelligent techniques applied in ITSs are classified into three groups, i.e., sequencing of curriculum, solving problems by a support system during man–machine interactions and providing adequate solution to learners by intelligent

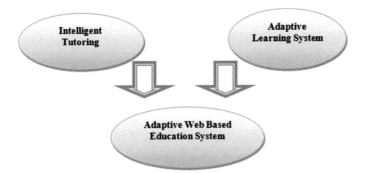

Fig. 2.2 Origin of adaptive web based E-learning technologies

analysis. The first and third techniques are in use since long time and categorized as the best-suited technologies in the purview of ITSs. All these technologies are enabled to support in the "intelligent" work of the teacher which was missing in the traditional systems.

Adaptive and Intelligent Web-Based Educational Systems (AIWBES) provides an option to move on real-time online system from the traditional one [7]. The development of web-based educational courseware made the task easy [3]. This system attempted to develop more adaptation by building a model comprising of the goals, preferences, and knowledge intelligently for each of the learners and provided learning as per needs arising out during interactions. Academician may get confused in both the technologies: intelligent tutoring system and Adaptive and Intelligent Web-Based Educational Systems (AIWBES).

Let them divide into three categories:

1. Intelligent but not adaptive (no student model)
2. Adaptive but not intelligent (not worthful)
3. Intelligent and adaptive (need of the hour).

- **Technology Fusion**

  Figure 2.3 shows the overlapping between IES and AES.

Fig. 2.3 Technology fusion of three systems

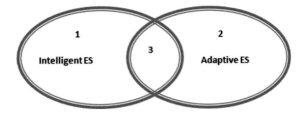

2.5   Adaptive E-Learning in Terms of Data Mining

In adaptive e-learning, there are numerous possibilities where it confirms users' behavior not only based on explicit entities like test results, but also based on implicit activities during interactions. The activities and their correlations in the continuous duplex communication and learning in a group vis-a-vis the actual learning performance during the entire learning process are very important aspects to critically analyze and find the outcomes. Therefore, e-learning is a potent platform having highly rated application environment for facilitating all types of computational intelligences (CSs) within the adaptive system [8]. Data mining techniques are used to explore all types of patterns which correlate with learners' behaviors during e-learning courses.

2.6   Application Area of Adaptive E-Learning

The areas in which adaptive e-learning techniques are frequently applied are given as below:

1. Dealing with the outcomes of learners' learning performance [9].
2. Course adaptability and corresponding learning suggestions based on the learners' learning behavior [10].
3. Evaluation and development of learning materials and educational web courses on continuous basis [11].
4. Feedbacks to tutors and learners of e-learning courses based on the learners' behaviors and tutor's analysis during learning process [12].
5. Developments for the detection, analysis, and evaluation of a typical learners' learning behavior [13].
6. Web usage mining and clustering in distance learning [14], i.e., discovering aggregate and individual path for learners and identifying virtual knowledge structure.
7. Data mining techniques used to build automated and inductive learner model as well as used to identify the background knowledge which are prerequisite to support system [15].

2.7   Adaptive Parameters in E-Learning?

Generally, a main question raises in our mind that what to be adapted? So the answer is adaptive ordering of educational courses, adaptive materials representation, adaptive ordering of selected learning objects, and adaptive kinds of flow of information from system to learners according to their needs, knowledge, interests, individual traits, etc., are considered as adaptive parameters in e-learning [1].

Adaptive e-learning system helps students in getting adaptive materials, ensures adaptive interactions, guides adaptive navigations, facilitates adaptive sequencing, and provides adaptive testing for self-assessment. Learners communicate and collaborate among peer group.

Adaptability is the process and capability of the e-learning system to provide personalization, i.e., contents or learning paths. It works with the help of a model having predefined rules based on learner's different measurable traits. It is a mechanism which depends upon different parameters and which make it successful. The different parameters responsible for adaptation are:

- **Cognitive Skills (CS)** by which humans acquire, map, process information, convert them into knowledge domain, and use them precisely in tests and other recognizable pursuits. Different researchers identify different aspects of CS [5]. It refers to user's own style of processing information. Cognitive abilities are inherent qualities that allow human brain to acquire, recognize and perceive information, convert them into meaningful discourse, convert information into knowledge accumulation and finally to use them for worthy pursuits to understand simple to complex human behavioral aspects [16].
- **Learning Style (LS)** is the way learners receive, acquire, interact, perceive, assume, act, think, and learn to get the knowledge. There are many studies on the classification of LS [17]. Brusilovsky [17] noticed that it was difficult to choose specific LSs for building a better learner modeling.
- **Subject Knowledge and Competency Levels** of the learner are considered prerequisite for better learning [16]. Learners coming from different background may compete differently and acquire knowledge and competency levels differently.

2.8 Adaptive LMS System Functionality (SF)

Adaptive learning systems (ALSs) are a new form of ITSs which use upgraded technology, latest learning theories, and basic psychological insights to develop instructional systems which adapt the things according to visibility, look, substance, material delivery process, and examining the knowledge levels and personal traits of the learner [18]. For Chieu [19], adaptability is the capability of a learning system which facilitates each learner to attain knowledge and its transformation under the conditioned environment. The ALS system works in tandem with the learners' needs, pre- and post-proficiency levels, differentiating learning styles, targeted goals, and preferences of the learners. The system is made capable of adjusting the learning materials and instructional recommendations as per learner's requirements similar to the traditional education system [20].

2.9 Learning Management System

Learning management systems in e-learning does mainly two kinds of measurable activities: One is related to communication and collaboration activities with the help of available tools like e-mail, chat/forums, video conferences, online blogs, etc., and another one is exploration activities linked with navigational data [21].

The main parameters of e-learning in LMS are:

(i)   the learner's personal profile, educational and background data
(ii)   the learner's shown preferences
(iii) the learning results and outcomes in SA/Test/Exams
(iv)   the learning contents adapted as per learner's preferences during learning process
(v)   the log data generated during man–machine interactions over the learning objects and with the system tutor
(vi)   the ordering relationships among the learning objects and their sequencing in terms of what, when, and how to provide the materials and
(vii) the specialized learning path for each of learner taking prominent characteristics of LSs and other factors.

2.10 Adaptivity in Learning Management Systems

Adaptivity defines in way of all sorts of automatic adaptation made available to the individual learner as per ones preferences [22]. In general, there are three kinds of adaptation [16] as given in next section.

2.11 Kinds of Adaptivity

- Adaptive Presentation: presenting the learning materials after adaptation to the learner as per ones identified LSs and preferences (mode of learning objects (Los)).
- Adaptive Navigation: adaptation of the learners' navigational path by providing personalized learning path (sequencing if Los).
- Adaptive Content: adaptation of the available learning materials according to the learner's profile, identified behavior, and their choice during learning process (types of Los).

2.12 Why Adaptation is Required?

- In traditional system, education process takes place when teachers and students interact face to face at same time same place. The learning activities are limited to the extent of what teacher has prepared learning material before the start of the learning session. This system does not care individual learning need and hence no adaptivity takes place. Therefore, the system fails to fulfil the prime objective of ensuring efficient learning process. Each learner has his own individual needs and characteristics.

- Most of LMSs do not consider learners' needs and preferences so a system must be required for addressing such issues and providing adaptive courses to the learners. Personalization refers to the fact that teacher knows everything about the learners in a traditional system and that is so in computerized environment system tutor knows the data related to appearance, behavior traits, etc. and provides adaptive learning experiences.
- The existing adaptive systems support adaptivity. But they only support limited adaptation in web-based education, and the course materials also do not change dynamically as expected. The materials are not reusable. In the existing system, dynamic personalization does not occur because of same learning materials are loaded in learning objects. Peer-level interaction is very low which is also required to enhanced learning process "to give the feel of classroom where learners informally interact with one another."
- In contrast, LMSs support tutor/system/instructors and enable the system to make online teaching comfortably. Rather, it should be learner centric. Adaptive LMS focuses a learner's needs.
- To gauge the challenges in future educational environment: improvised e-learning required to enable students free from constraints of classroom and time, to fulfill the requirement of anytime-anywhere learning, to outreach a large number of students and support their preferred mode of learning, support the concept of learning by doing or earning with learning, to harness maximum of its cost-effectiveness, to minimize the requirement of huge infrastructure, to help in scenario of inadequate student–teacher ratio, etc.
- As the education system moves from print to digital classrooms, learning products or objects change rapidly and academic institutions and policies must respond accordingly to keep things updated.
- Drawbacks of present e-learning system: Static content, design based on "one size fits all." In this system, teacher and learner need not required to present at same time same place. Here, the learning process takes place with the help of websites, emails, blogs, etc.
- It is vital for tutors to adopt adequate teaching strategies to effectively meet out the preferences of individual learner. Preferred modes of output vary from the variation of the input data and from individual to individual; only an adaptive e-learning system can cater the needs in such situations.
- Lack of ability to adapt content to individual student having different interests, knowledge background, learning style, etc. in e-learning environment.
- Most of students unable to channelize their energy in proper domain due to the lack of personalized adaptive e-learning framework.

- Even excellent teacher cannot guarantee student's success. It equally depends upon student's individuality which is considerably addressed in adaptive e-learning.
- Students may get puzzled which course is the best to choose without any recommender system.

2.13 Personalization with Adaptivity

Adaptive e-learning system may constitute hybrid approach (containing both data and literature) that can automatically identify learners' preferences and learning style of learners by applying artificial intelligence with corresponding soft computing tools and finally offering the personalized course material according to the identified learner's learning style of preferences. The data can be mined distinctly related to learners' behavior during their interaction with the system through learning objects and other collaborative activities. Mapping rules help to decipher hidden patterns from the available data. The patterns are classified into different learning styles based on different learning style models, e.g., Felder–Silverman learning style model (FSLSM). Data can be acquired in web-based e-learning system say, i.e., Adaptive Personalized Intelligent e-learning System (APIE LMS) based on Moodle platform. The data associated with the learning objects and other collaborative activities can be analyzed in terms of the number of visits (frequency), time that learner spent on learning objects (time expected to time consumed), course completion percentage, order of selecting learning materials/objects, and other collaborative activity-related factors.

Moreover, many researchers postulated that adopting personalization based on learning styles, motivation, and other related factors in adaptive e-learning makes learning easier for learners and increases their knowledge level. It has been emphasized that the adaptive e-learning system using EDM techniques helps students bridging gaps related to knowledge, skills, aptitudes, and preferences. However, the effectiveness of e-learning system can be maximized only by the fact that learner uses the system coherently at desired levels.

2.14 Adaptive E-Learning Its Scope and Challenges

Adaptivity refers to timely intervention, delivery of most suited learning materials, upgrading of the learning process, and improving the evaluation system. It is very important that teacher uses a range of teaching planning, scheduling, strategies, and methodologies precisely to meet the quests of individual learners [23].

2.15 Scope—A Tool to Manage Shortcomings of E-Learning

Despite the technical progress of IT and telecom technologies in almost every sphere of life, e-learning is not popularized as much as it is needed. Some people still consider it as sub-standard in education domain. But it is not the case in present scenario. The practical experiences of a study is discussed in [24] that how adaptive e-learning was introduced in the higher education and how it helped learners to get better result. It showed that adaptivity successfully led to overcome the shortcomings of the e-learning system.

2.16 Some Existing Challenges

Improvising adaptive e-learning systems and its proper implementation into the educational processes are indeed complex and expensive mechanism that requires a high level of involvement from all stakeholders in the process. Some of the challenges are discussed here as follows:

- Such systems are formulated for the objective of ensuring adaptivity in educational domain. However, in practical scenario, the learners do not follow the chalk out plan, show odd behavior like frequently log-in or log-out and thus, hinder learning process.
- It is cumbersome in terms of recreating and developing new L.M intermittently. The research has shown that systems cannot exchange either resources or learner's data or are most often remain in fix with limited data applications and experimental clarifications.
- To implement effective adaptive system, its basic ingredients like course administration, creating and developing materials, etc., are required to be of high standard.
- As the entire system is very complex, the learners should have some prerequisite knowledge about the system. So that, the learning process would have some ease. In absence of such knowledge, the learning process will not fetch desired result and it will become void ab initio.
- A large part of communication and social interactions between the stakeholders in the e-learning system is exchanged through e-mail, Twitter, Facebook, chat, forum and video conference. But the physical or system interruptions hinder learning process largely.
- In most of the courses, the duration of the course is an important factor. It is very much required to learn about the learners' needs during learning process and their time limits and then provide adaptation accordingly to complete the course within that time. In the process, prerequisite knowledge, prior experiences, intention to study well and fetch high marks, competitive mindset during learning, and exploit the opportunities at the maximum levels are factors behind successful adaptation.

The existing AWBES system provides adaptation, but it does not guarantee its successful completion within the given time slot.

- The concerns of the adaptivity do not merely linked with adaption, but it equally depends upon framework, design of modules, process flow strategies, and smooth implementation of entire process. Moreover, the LMSs do not offer adequate level of adapting materials and processes [25].

Although the adaptivity concept is simple, however it is very challenging to develop a framework. The challenges lie in the complexity of capturing data, representing, and processing all valuable information. The other challenges that motivated many research scholars to the use A.E.L. are as given below:

(A) Scalability [26], (B) multidimensionality, and (C) heterogeneous [12].

2.17   Process of Adaptation

This book addresses formulation of a framework which provides adaptive contents to learners in view of their learning preferences, which is an important factor for the efficiency and effectiveness of the learning process. A framework is discussed in Chap. 5 about adaptive e-learning system through which system can provide best-suited personalized contents based on input data and their scientific patterns analyzed by the data mining techniques. This framework is used to develop a learner model to identify learning style preferences of learner by applying methods based on fuzzy cognitive map (FCM) techniques in Chap. 6. It is a fusion of soft computing techniques of fuzzy logics and neural networks.

References

1. Premlatha KR, Geetha TV (2015) Learning content design and learner adaptation for adaptive e-learning environment: a survey. Artif Intell Rev. https://doi.org/10.1007/s10462-015-9432-z
2. Saini DK, Prakash LS, Goyal M (2012) Emerging information technology and contemporary challenging R & D problems in the area of learning: an artificial intelligence approach. In: AICERA 2012—annual international conference on emerging research areas innovative practices and future trends. https://doi.org/10.1109/AICERA.2012.6306748
3. Brusilovsky P (2004) Adaptive educational hypermedia: from generation to generation. In: Proceedings of 4th Hellenic conference on information and communication technologies in education, Athens Greece, pp 19–33, 2004, [Online]. Available: https://www2.sis.pitt.edu/~peterb/papers/PB_ETPE_04.pdf
4. Sakurai Y, Takada K, Tsuruta S, Knauf R (2012) A case study on using data mining for university curricula. In: 2012 IEEE international conference on systems, man and cybernetics (SMC), pp 3–4. https://doi.org/10.1109/ICALT.2012.212
5. Desmarais MC, Baker RSJD (2012) A review of recent advances in learner and skill modeling in intelligent learning environments. User Model User-Adapted Interact 22(1–2):9–38. https://doi.org/10.1007/s11257-011-9106-8

6. Lin CF, Yeh YC, Hung YH, Chang RI (2013) Data mining for providing a personalized learning path in creativity: an application of decision trees. Comput Educ 68:199–210. https://doi.org/10.1016/j.compedu.2013.05.009

7. Abraham G, Balasubramanian V, Saravanaguru RAK, Abraham G (2013) Adaptive e-learning environment using learning style recognition 2(1):23–31

8. Kock M (2008) Computational intelligence for communication and cooperation guidance in adaptive e-learning systems. In: 16th Work Adapt User Model Interact Syst, pp 32–34

9. Dias SB, Hadjileontiadou SJ, Hadjileontiadis LJ, Diniz JA (2015) Fuzzy cognitive mapping of LMS users' quality of interaction within higher education blended-learning environment. Expert Syst Appl 42(21):7399–7423. https://doi.org/10.1016/j.eswa.2015.05.048

10. Sweta S, Lal K (2015) Proceedings of the fifth international conference on fuzzy and neuro computing (FANCCO-2015). In: Ravi V, Panigrahi KB, Das S, Suganthan NP (eds) Springer International Publishing, Cham, pp 353–363

11. Romero C, Ventura S (2010) Educational data mining: a review of the state of the art. IEEE Trans Syst Man Cybern Part C (Applications Rev) 40(6):601–618. https://doi.org/10.1109/TSMCC.2010.2053532

12. Castro F, Vellido A, Nebot À, Mugica F (2007) applying data mining techniques to e-learning problems\r. Stud Comput Intell 62:183–221

13. Dascalu M-I, Bodea C-N, Moldoveanu A, Mohora A, Lytras M, de Pablos PO (2015) A recommender agent based on learning styles for better virtual collaborative learning experiences. Comput Human Behav 45:243–253. https://doi.org/10.1016/j.chb.2014.12.027

14. Romero C, Pechenizkiy M, Calders T, Viola SR (2007) international workshop on applying data mining in e-learning (ADML' 07) as part of the second European conference on technology enhanced learning (EC-TEL07)

15. Agudo-Peregrina ÁF, Iglesias-Pradas S, Conde-González MÁ, Hernández-García Á (2014) Can we predict success from log data in VLEs? Classification of interactions for learning analytics and their relation with performance in VLE-supported F2F and online learning. Comput Human Behav 31(1):542–550. https://doi.org/10.1016/j.chb.2013.05.031

16. Dominic M, Xavier BA, Francis S (2015) A framework to formulate adaptivity for adaptive e-learning system using user response theory. IJ Mod Educ Comput Sci 1:23–30. https://doi.org/10.5815/ijmecs.2015.01.04

17. Hsiao I, Bakalov F, Brusilovsky P, König-ries B (2013) Progressor: social navigation support through open social student progressor: social navigation support through open social student modeling, 1(412)

18. Millán PB (2007) User models for adaptive hypermedia and adaptive educational systems. In: The adaptive web, pp 3–53

19. Chieu V, Deville Y (2005) Constructivist learning: an operational approach for designing adaptive learning environments supporting cognitive flexibility. Louvain Fac

20. Mohammad Bagheri M (2015) Intelligent and adaptive tutoring systems: how to integrate learners. Int J Educ 7(2):1. https://doi.org/10.5296/ije.v7i2.7079

21. Caputi V, Garrido A (2015) Student-oriented planning of e-learning contents for Moodle. J Netw Comput Appl 53:115–127. https://doi.org/10.1016/j.jnca.2015.04.001

22. Özyurt Ö, Özyurt H, Baki A (2013) Design and development of an innovative individualized adaptive and intelligent e-learning system for teaching–learning of probability unit: details of UZWEBMAT. Expert Syst Appl 40(8):2914–2940. https://doi.org/10.1016/j.eswa.2012.12.008

23. Zhang D, Nunamaker JF (2003) Powering e-learning in the new millennium: an overview of e-learning and enabling technology. Inf Syst Front 5(2):207–218. https://doi.org/10.1023/A:1022609809036

24. Karagiannis I, Satratzemi M (2014) Comparing LMS and AEHS: challenges for improvement with exploitation of data mining. In: 2014 IEEE 14th international conference on advanced learning technologies, pp 65–66, 2014. https://doi.org/10.1109/ICALT.2014.29

25. Sciences O, Ili J (2012) Providing adaptivity in Moodle LMS courses. In: Krčo S (ed) Adaptive e-learning systems, vol 15, pp 326–338
26. Maiorana F, Mongioj A, Vaccalluzzo M (2012) A data mining e-learning tool: description and case study. Proc. World ..., vol I, pp 6–9. https://www.iaeng.org/publication/WCE2012/WCE 2012_pp432-435.pdf

Chapter 3
Educational Data Mining Techniques with Modern Approach

3.1 Introduction

This book covers mainly two broad perspectives: one is educational data mining (Technological Aspects) and another one is Personalized Adaptive e-Learning System (Application Aspects). There are wide ranges of literatures available related to adaptive e-learning in educational data mining. Here, we discussed the approaches used to investigate adaptive e-learning system in measurable units and extract valuable information in terms of Personalized Adaptive e-Learning.

This chapter aimed to discuss multifaceted and updated snapshots of the present state of researches. The data mining techniques were predominantly used in adaptive e-learning system in which learning style diagnosis and learning ability factors like motivation and knowledge ability factors were discussed. Moreover, a study of data mining in e-learning had been explored and adaptive e-learning had also been scanned. It covers the classification problems and related solutions, clustering and other data mining techniques, discussions and opportunities for using of data mining and optimized key e-learning resources [1].

3.2 Data Mining in Terms of Adaptive E-Learning and Web Personalization

Data mining techniques are nothing but extraction and unwinding of hidden predictive patterns which culminate into useful outputs from a huge database. They are used by researchers, academicians, and administrators in various knowledge discoveries, commercial, and other applications like education system, adaptive e-learning bioinformatics, and e-commerce [2]. Adaptivity and web personalization have added advantages [3].

© The Author(s), under exclusive license to Springer Nature Singapore Pte Ltd. 2021
S. Sweta, *Modern Approach to Educational Data Mining and Its Applications*,
SpringerBriefs in Computational Intelligence,
https://doi.org/10.1007/978-981-33-4681-9_3

3.2.1   Soft Computing Techniques in Data Mining

It is a well-defined fact that words are less precise than numbers. But in human analysis, numbers have also its limitations. In this regard, high precision in words is sought for and indeed it is order of the day now to use it extensively [4]. Other way round, it carries high cost considerably. Cost-precision optimization necessitates and leads to the uses of soft computing techniques. In fact, the principle of soft computing judiciously used to leverage the uncertainty and precision to reduce the cost and enhance the robustness of the system [5]. In terms of analysis of human behavior and preferences, it is important to use texts because the extracted information is not enough to use numerals and it will not lead to any conclusion. This is usual phenomena why soft computing is used, i.e. when "hard" computing fails to produce any solution for the cases dealt with the complex problems associated with human decision and thoughts. Soft computing is an innovative approach which constructs computationally intelligent system. Such intelligent system is enabled to possess and exhibit human like expertise within a given domain. It becomes adaptive itself according to the variable factors, learn to do better in dynamic environment and precisely explains how it makes decision or acts in the desired order. Here, it is pertinent to mention that the role model for soft computing is human mind and its similar functional aspects. Such human like decision-making system confers their utility in terms of solidarity, understanding steps in decision making and cost optimization [6].

Soft computing is a group of mathematical, statistical, logical, and graphical techniques used to get baneful information from heaps of large database. Soft computing techniques work as tool for artificial intelligence. They are generally categorized into fuzzy logics, neural networks, genetic algorithms, rough sets, etc. The techniques process meaningful real-life cases and provide worthy information necessary for taking suitable decisions. They may work in tandem complementarily, synergistically, and flexibly among themselves as per data and desired outcomes. They are targeted to validate the tolerance level of data set to find precision values, trace uncertainty, apply logics to approximate results, understand the meaning of partial truth. This is a summary on soft computing used for different data mining tasks. As of now, many data mining tools have been enunciated commercially based on soft computing techniques.

Soft computing DM techniques were used as fuzzy logic methods [7], Bayesian network [8], artificial neural networks [6] and evolutionary computation [9], graphs and trees [10], genetic algorithm, hybrid intelligent system, etc.

3.2.2   Advantages of Soft Computing

Of late, data mining is a latest, burgeoning, and significant area of research. It became very popular in recent past and now it comes as interdisciplinary areas of data science, big data, data analytics, and many more as nature of data get changed. Simultaneously, soft computing tools as data mining techniques are also emerged as front-runner for analyzing data and decipher hidden patterns like human brain. Among various techniques used in DM, the soft computing methods are very apt for resolving the different issues because of their qualities of solidarity, enabler of adaptivity, capable of dual processing, possess high precision value, and show high degree of tolerance. They were used in data mining which produced various analytical and computing techniques like ANN, GA, and FL [11]. They widely used approximation value or rough value sets which enhanced the efficiency of data mining techniques substantially. The result showed genesis of a more intelligent and robust system over the traditional techniques in which the human interpretability, low-cost analysis, and approximate solution were identified as critical factor. The traditional quantitative techniques used in statistics and economics do not measure data because of its criticality in streamlining the system's outputs into quantitative functions. Therefore, the application of soft computing in data mining is most sought for large database. Similarly, the characteristic of neural networks can be used to identify and establish functions between numerous variables.

Soft computing techniques mainly comprise of:

1. Fuzzy system
2. Neural networks
3. Genetic algorithm
4. Hybrid intelligent system.

3.3 Soft Computing Techniques in Personalized Adaptive E-Learning System

Learners' performance scaling and analysis can be carried out in intelligent tutoring system under the framework of Neuro-Fuzzy concepts in which fuzzy logics can be used [12].

Artificial neural networks (ANN) and evolutionary computation (EC) concepts are introduced, and its related designs are examined. Artificial neural networks are best suited for developing navigational system because they are coupled with Multilayer Perceptron techniques. This concept is widely used.

The learners' learning behaviors depend upon many factors. These factors are called combination of multiple classifiers (CMC) [1]. The evaluation of learning behaviors is done by applying evolutionary algorithms.

Learners' comprehensibility, their profile and behavioral classifications and their results or grades predictability can be obtained by an educational web-based system in which random code generation and mutation processes can be applied on the logged data [1].

Table 3.1 Learning style incorporated for adaptive systems

	ES	FS	NN	GA
Knowledge representation	rather good	good	bad	rather bad
Uncertainty tolerance	rather good	good	good	good
Imprecision tolerance	bad	good	good	good
Adaptability	bad	rather bad	good	good
Learning ability	bad	bad	good	good
Explanation ability	good	good	bad	rather bad
Knowledge discovery and data mining	bad	rather bad	good	rather good
Maintainability	bad	rather good	good	rather good

* The terms used for grading are:

□ - bad, ▪ - rather bad, ▤ - rather good and ▦ good

3.3.1 Comparative Analysis of Expert Systems, Fuzzy Systems, Neural Networks, and Genetic Algorithms

Table 3.1 shows the soft computing techniques and their qualifying parameters.

3.3.2 Intelligent Hybrid System

- A hybrid intelligent system is a comprehensive and target-oriented system. It is designed to overcome the problems related to an individual system by adding the concepts of two or more intelligent systems. For example, a hybrid Neuro-Fuzzy system is formed by fusion of neural networks and fuzzy system.
- The fusion of components of the added systems is formulated in the right perspectives to develop a good hybrid system.
- The different soft computing tools are used in such a way as complementary to one another. Their superimposition does not provide impedance. It enhances system's comprehensibility and robustness [13], for example, Neuro-Fuzzy system, Neuro-Fuzzy Genetic, Fuzzy Genetic, and Neuro-Genetic.

3.3.3 Neuro-Fuzzy Approaches

Fuzzy modeling based on IF-THEN rules is well suited for inference systems to overcome the problem of transparency and transferability of result to observe network behavior. Analysis of rule definition, finding valuable information from data after applying rules and finally providing adaptation demand highest level of human intervention [14]. Fuzzy logics coupled with ANNs take strengths of respective computing techniques to surmount the individual's weaknesses. They are used to detect hidden information more comfortably. The combined approach can be implemented by applying the fuzzy inference engine from the fuzzy system and integrated it with the ANN. ANN is capable to enable learner to learn in a better way to achieve desired adaptivity.

3.3.4 Advantages of Combination of Neuro-Fuzzy

- Neural networks recognize patterns easily, but do not explain how decision was arrived; i.e. pattern analysis does not lead to making decisions.
- Fuzzy logics explains better all about decision-making mechanism, but unable to propose new rules by their own system. Improper information hinders in this regard.
- Neural networks decipher patterns and enable the system to adapt according to changes in the environment, whereas fuzzy inference system incorporates human knowledge and shows distinct decision-making mechanism. So, they work complementary to each other.
- The limited usability of individual systems led to formation of hybrid intelligent system. The hybrid system constructs better intelligent system.
- In Neuro-Fuzzy hybrid system, the domain knowledge data can be changed into linguistic variables and then fuzzy rules can be applied.
- A Neuro-Fuzzy system is a set of IF-THEN rules. It reduces dependency on expert knowledge while formulating an intelligent system [14].

3.4 Fuzzy Cognitive Map (FCM)

In the last couple of years, there is tremendous growth in fuzzy logics in terms of their uses and applications across various studies. They are very useful in education like analyzing the performance of students [15].

Fuzzy cognitive mapping [FCM] is the structured of most relevant concepts, their casual interlinked strengths where concept values change dynamically. The fuzzy logics and neural networks are applied intelligently in FCM to understand the casual concepts more precisely for detecting hidden patterns. They are used for pictorial and graphical presentations. They are useful in formulating a model and understanding

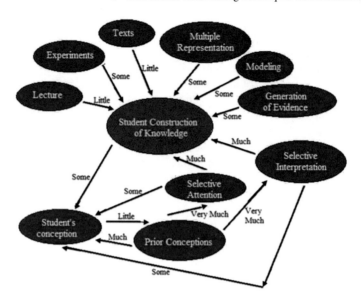

Fig. 3.1   Fuzzy cognitive map in education system

soft knowledge areas like international affairs, hospitability, politics, and education [15].

It is required to understand the theoretical aspects of cognitive mapping when FCMs applied in education [16]. Cognitive concepts added with fuzzy rules strengthen FCMs. It is a system depicting information and conceptual knowledge about the learning object graphically. It is represented by circles for core and interlinked concepts and arrows for the interconnected strength between them.

3.4.1   Application of Fuzzy Logic in Education

Education is a soft skill knowledge area in which the characteristics of learners and learning process are suitably measured in terms of values in a range as shown in Fig. 3.1. Here, it is not advisable to use the cues in yes or no format. It is better to define in terms of membership functions. Therefore, fuzzy logic is required.

3.4.2   Advantages and Disadvantages of Fuzzy Logic

- **Advantages**:
    - Accept fuzzy, imprecise, data with some error or contradictory variables
    - Flexibility in reconciling the conflicting input /output variables

- Fuzzy rules easily changed as and when required
- Relates input to output in linguistic terms
- Easy to get information or understand
- Cost effective because of simple design
- Robust system because of linguistic flexibility
- Enable simple knowledge capturing tools and methodologies
- Support, FCMs in capturing more information related to concepts and weights between the concepts [17].
- Supports FCMs in dynamic, combinable, and tunable environment.
- Supports FCMs to express hidden information.

- **Disadvantages**:

  - Many scholars have proposed many rules in this domain. It is difficult to follow the rules as they are proposed with certain limitations about usages, input, or output variables.
  - Giving same importance to all factors of the fuzzy logics causes difficulty in analyzing when certain sets of data become more important than others.

3.5 Data-Driven Approach Versus Literature-Based Approach

It is prerequisite for enabling system for auto-detecting the LSs for the required adaptation. Index of Learning Style (ILS) Questionnaires of FSLSM can be used for detecting learning style of the learners. In which, if learners select wrong answers intentional or unintentional, it will mislead the result in ILS questionnaire. The new approaches are meant to overcome the disadvantages of ILS, where both approaches can be used distinctly. Figure 3.2 shows where both the approach applied.

In data-driven approaches, two methods are used—Bayesian network [8] and hidden Markov models, and decision trees [18]. Whereas literature-based approach uses questionnaire, and the information extracted from system was first done by Graf et al. [19]. They undertook different data set and specified information in the detection of LS.

3.5.1 The Data-Driven Approach

The approach deals in real data which provide more accurate results in findings as studied by Garcia et al. [20]. However, data accuracy is limited to the characteristics of data. The collection of good data set is very much desirable, and it is always a challenging task because most of the available data are found in scattered states.

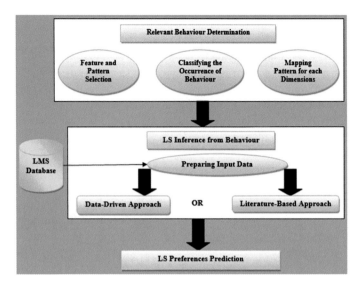

Fig. 3.2 Data-driven versus literature-based

Another study was conducted by Cha et al. [18]. They examined the use of decision trees (DT) and hidden Markov models (HMM) for detecting learning styles based on FSLSM.

The learners' behavior can be observed during e-learning courses in an intelligent learning system. The learners can fill out the ILS questionnaire for evaluating the models and establishing the relation in models for comparing precision values. Moreover, the data related to human behavioral and interactions cannot be found in crisp values, and these can be measured in the defined ranges.

3.5.2 The Literature-Based Approach

This approach is similar to ILS methodologies for evaluating learning styles [19]. It gives better result for the generic and applicable data which are collected and analyzed for the entire learning process in the e-learning course. It is comparatively a new methodology used in contemporary studies more interlinked to human process analysis. However, this approach faces some problems like prioritizing the significance of the data. On the contrary, it failed to extract right inferences and cues while calculating LSs.

A similar method was presented by Graf et al. using this approach where LSs identification was based on FSLSM [21].

New model discussed in this book is structured on data-driven, but it has the added advantages of literature-based approach like concept of linguistic variables. These linguistic variables can be implemented suitably to understand soft knowledge domain and more precisely. So, the discussed model is presumed as hybrid one.

3.6   Learner Modeling Techniques for Personalized and Adaptive E-Learning

One most prominent distinguishable properties of an adaptive system are moving away from teacher-centric model to user-centric. The learner model contains all relevant information about an individual learner and how adaption works [22] and how soft computing tools are used by taking into account of all measurable behavioral entities of learners including their interactions with the system at the learning objects.

The users can behave differently and exhibit different activities during the learning processes [23]. Adaptation of learning styles can be analyzed, and content adaptation can be provided in an attempt to match learner's characteristics to the specific LSs.

Literature survey can be conducted to find out different models and methodologies used in the past for creating learner's model. Fuzzy logics and neural networks are most frequently used techniques in the learner modeling [24], because of their abilities to learn from uncertainty and incomplete or missing patterns of learner's behavior.

Some important learner modeling techniques used in adaptive systems are given in succeeding subsections.

3.6.1   Bayesian Belief Network

BBN is used to provide learning path in the specified learning style models. It provides many learning paths for a learner. Optimization of the learning path is the most important aspect in detecting the LS [8]. It is a directed graph in which each node presents an uncertainty variable. The edges having some values are causal links between the variables. The value of each node is associated with a probability table. The probability table consist values based on conditional probability that are calculated from the casual relationship between the node and its parent [25].

It can help in constructing the model by applying its feature to understand machine learning. It is a unique modeling technique called feature-based modeling that is used in the tutoring systems. It can also be used to construct behavior-oriented learner model.

Limitations

The machine learning generates large amount of uncontrollable data, e.g. browse, log in /log out frequencies, clicks, streams, search logs, etc. The author of this book suggests building an integrated system for providing desired adaptation. The relevant data of human behavior and some assumptions during the course can be suitably observed. These data are fuzzy in nature. So, probability versus fuzziness is to be focused as it defines limitations of BBN.

3.6.2 Fuzzy Logic-Based Technique

The limitation of traditional machine learning system can be overcome by soft computing techniques for creating learner model. Fuzzy logic is one of soft computing techniques that capture the real-life data with acceptable level of ambiguity. It constructs a typical model that categorizes learners with the help of membership functions and accordingly provides recommendations [26].

Limitations

System cannot decipher information by merely applying fuzzy logics. Fuzzy logic along with the machine learning is required to capture measurable characteristics about the learner's behavior.

3.6.3 Neural Network-Based Techniques

The valid analysis and meaningful patterns disclosures are obtained from a pool of huge, imperfect, and imprecise data by applying techniques of neural networks [27]. They deal efficiently in the domain of human behavior analysis and constructing-related models [1], modeling learners' behavior in intelligent tutoring systems and very useful in doing classification, finding navigational path and recommending accordingly.

Limitations

The above system does not support two phase learning algorithms where neural networks and machine learning techniques are used to construct learner's model for finding learning needs. Huge amount of data are applied to find the requirements in which NN does not support.

3.6.4 Fuzzy Clustering-Based Techniques

There are mainly two types of clustering—one is hard clustering or non-fuzzy clustering and another one is soft clustering or fuzzy clustering [28]. Fuzzy clustering is applied in measuring the knowledge level of the learner in concept module where performance in tests is not of much use. Such clustering fuzzifies the learners like the human analysis.

3.7 Learning Style-Based Individualized Adaptive E-Learning

The author thoroughly studied, reviewed, and analyzed 120 odd publications from 2005 to 2014 and reviewed various papers in detail. The summaries are analyzed properly and concepts are documented accordingly. This book discusses the existing trends, analyses of experimental outcomes and findings about model's effectiveness. It is also discussed in detail all about the future scope for minimizing learning gaps.

The studies on developing learning systems and preferred LSs of the learners have taken center stage in the last couple of years. It is gauged up as one of the prominent factors in evaluating individuality. Gradually, the traditional system is suitably replaced by adaptive ones. The new system is based on innovation and precisely performs as per learners' needs.

Changes in teaching and learning methodologies enhance efficiency in learning process and effectiveness of e-learning system. They can minimize gaps in terms of "what is asked and what is provided" [29]. The learning differences and learning needs can be studied well to provide learner's friendly environment [23] what is required most. Another study discussed the Integration of Information and Communication Technology (ICT) into educational environment and its important contributions in the learning processes. The adaptive and adaptability are buzz words in personalized and soft learning domains [30]. The important aspects can be addressed before making adaptive e-learning systems "what to be taken as basis for adaptation," "what are the available resources," and "how they are utilized to achieve the targeted adaptivity."

Learning style is vital in considering individual differences while formulating adaptive e-learning environments [31]. It is deduced as different learning choices and different personality traits (Dunn & Dunn, 1978; Felder & Silverman, 1988; Veznedarog ˇlu & Özgür, 2005). Brusilovsky [32] described general architectures, models, and taxonomies referring AEHs. The content analysis studies in view of technological support to build learning environments are given by [33, 34].The three prominent studies were combined in a system called WELSA proposed by Popescu [35], provides another insight.

Of late, a large number of adaptive e-learning systems are developed by researchers. Many of them are very popular. They have incorporated the basics of

Table 3.2 Learning style incorporated for adaptive systems

System	Learning style
ARTHUR et al. [37]	Text styles, spoken lectures, and visual-interactive
iWeaver et al. [38]	Spontaneous, experiment loving, kinesthetic, holistic, reflective, oratory, visual, analytical styles.- Dunn and Dunn LS model
CS388 [39]	Four dimensions excluding Active/Reflective Learning Styles: FSLSM
AEC-ES [40]	Field-Dependent (FD) and Field-Independent (FI) LSs
LSAS [41]	Global-Sequential dimension: FSLSM
MANIC [42]	Graphic Versus Textual information
INSPIRE [43]	Honey and Mumford Classified activists, pragmatists, reflectors, and theorists - based on Kolb's LS
Tangow [44]	Sensing-Intuitive dimension: FSLSM
Lee [45]	Symbolic, Iconic, Enactive, etc.

learning styles based on the notion of matching the learning strategies with the learning styles of the individuals to improve learner performance [36].

The adaptation carried out with the help of different media representations for each of the learner given in ARTHUR [37], iWeaver [38], CS388 [39], and MANIC. ARTHUR and iWeaver are represented by very similar choices of learning style representations.

CS388 includes graphs, movies, texts, and slide shows. Similarly, MANIC used graphic and textual information. In Learning Styles Adaptive System (LSAS) [41], the sequential learners are assigned with advanced materials, passing maximum instructions, receiving feedbacks, and more structured lessons. In the contemporary studies led by Tangow [44] and INSPIRE [43], adaptation presents different sequences of alternative content and the concepts used in e-learning system.

Many of the proposed systems except iWeaver and MANIC diagnosed the learning styles with the help of psychometric tests. Discussed adaptability in case of an inactive, iconic and symbolic learning styles by Lee & Kim [45] applying EIS Theory to an adaptive learning system. Table 3.2 illustrates learning style can be incorporated into adaptive e-learning.

References

1. Castro F, Vellido A, Nebot À, Mugica F (2007) Applying data mining techniques to e-learning problems\r. Stud Comput Intell 62(2007):183–221
2. Romero C, Ventura S (2007) Educational data mining: a survey from 1995 to 2005. Expert Syst Appl 33(1):135–146. https://doi.org/10.1016/j.eswa.2006.04.005
3. Sweta S, Lal K, "Adaptive e-Learning System: A State of Art". Int J Comput Appl 107(7):13–15 http://www.ijcaonline.org/archives/volume107/number7/18762-0046
4. Yadav RS, Singh VP (2011) "Modeling academic performance evaluation using soft computing techniques: a fuzzy logic approach". Int J Comput 3(2):676–686. doi:0975–3397

5. Peña-ayala A (2014) "Educational data mining: a survey and a data mining-based analysis of recent works". Expert Syst Appl 41(4) PART 1:1432–1462. https://doi.org/10.1016/j.eswa.2013.08.042

6. Azevedo Dorca F, Vieira Lima L, Aparecida Fernandes M, Roberto Lopes C (2012) Consistent evolution of student models by automatic detection of learning styles. IEEE Lat Am Trans 10(5):2150–2161. https://doi.org/10.1109/tla.2012.6362360

7. Doctor F, Iqbal R (2012) "An intelligent framework for monitoring student performance using fuzzy rule-based linguistic summarisation". IEEE Int Conf Fuzzy Syst, August. https://doi.org/10.1109/fuzz-ieee.2012.6251312

8. Carmona C, Castillo G, Millan E (2008) "Designing a dynamic bayesian network for modeling students' learning styles". 2008 Eighth IEEE Int Conf Adv Learn Technol, https://doi.org/10.1109/icalt.2008.116

9. Premlatha KR, Geetha TV (2015) "Learning content design and learner adaptation for adaptive e-learning environment: a survey". Artif Intell Rev, https://doi.org/10.1007/s10462-015-9432-z

10. Wu D, Zhang G, Lu J (2014) "A fuzzy tree matching-based personalised e-learning recommender system"

11. Abraham G, Balasubramanian V, Saravanaguru RaK (2013) "Adaptive e-learning environment using learning style recognition". 2(1):23–31

12. Caputi V, Garrido A (2015) Student-oriented planning of e-learning contents for moodle. J Netw Comput Appl 53:115–127. https://doi.org/10.1016/j.jnca.2015.04.001

13. Salehi M, Nakhai Kamalabadi I, Ghaznavi Ghoushchi MB (2013) "An effective recommendation framework for personal learning environments using a learner preference tree and a GA". IEEE Trans Learn Technol 6(4):350–363, https://doi.org/10.1109/tlt.2013.28

14. Stathacopoulou R, Magoulas GD, Grigoriadou M, Samarakou M (2005) Neuro-fuzzy knowledge processing in intelligent learning environments for improved student diagnosis. Inf Sci (Ny) 170(2–4):273–307. https://doi.org/10.1016/j.ins.2004.02.026

15. Cole JR, Persichitte Ka (2000) "Fuzzy cognitive mapping: applications in education". Int J Intell Syst 15:1–25, https://doi.org/10.1002/(sici)1098-111x(200001)15:1%3c1::aid-int1%3e3.0.co;2-v

16. Saxena N, Saxena KK (2010) "Fuzzy logic based students performance analysis model for educational institutions". January: pp 79–86

17. Chrysafiadi K, Virvou M (2014) "Fuzzy logic for adaptive instruction in an e-learning environment for computer programming". Fuzzy Syst IEEE Trans PP(99):1, https://doi.org/10.1109/tfuzz.2014.2310242

18. Cha HJ, Kim YS, Park SH, Yoon TB, Jung YM, Lee JH (2006) "Learning style diagnosis based on user interface behavior for the customization of learning interfaces in an intelligent tutoring system". 8th Int Conf Intell Tutoring Syst Lect Notes Comput Sci 4053:513–524, https://doi.org/10.1007/11774303_51

19. Graf S, Viola S, Kinshuk (2007) "Automatic student modelling for detecting learning style preferences in learning management systems". IADIS Int Conf Cogn Explor Learn Digit Age 1988:172–179, http://sgraf.athabascau.ca/publications/graf_viola_kinshuk_CELDA07.pdf

20. Garcı P, Schiaffino S, Campo M (2007) "Evaluating bayesian networks õ precision for detecting students õ learning styles". 49:794–808, https://doi.org/10.1016/j.compedu.2005.11.017

21. Baldiris S, Graf S, Fabregat R (2011) "Dynamic user modeling and adaptation based on learning styles for supporting semi-automatic generation of IMS learning design". 2011 IEEE 11th Int Conf Adv Learn Technol. pp 218–220, https://doi.org/10.1109/icalt.2011.70

22. Kobsa A (2001) Generic user modeling systems. User Model. User-adapt Interact 11(1–2):49–63

23. Millán PB (2007) "User models for adaptive hypermedia and adaptive educational systems". Adapt Web, pp 3–53

24. Hawkes LW, Derry SJ, Rundensteiner EA (1990) Individualized tutoring using an intelligent fuzzy temporal relational database. Int J Man Mach Stud 33(4):409–429

25. Sweta S (2015) "Adaptive and personalized intelligent learning interface (APIE-LMS) in e-learning system. 10(21):42488–42492

26. Georgiou DA, Makry D (2004) "A learner's style and profile recognition via fuzzy cognitive map". Proc IEEE Int Conf Adv Learn Technol ICALT 2004, pp 36–40, https://doi.org/10.1109/icalt.2004.1357370

27. Dagez HE, Baba MS (2008) "Applying neural network technology in qualitative research for extracting learning style to improve e-learning environment". 2008 Int Symp Inf Technol, pp 1–6, https://doi.org/10.1109/itsim.2008.4631550

28. Ross TJ (2004) *Fuzzy logic with engineering applications*

29. Özyurt Ö, Özyurt H (2015) Learning style based individualized adaptive e-learning environments: content analysis of the articles published from 2005 to 2014. Comput Human Behav 52:349–358. https://doi.org/10.1016/j.chb.2015.06.020

30. Özyurt Ö, Özyurt H, Baki A (2013) Design and development of an innovative individualized adaptive and intelligent e-learning system for teaching–learning of probability unit: details of UZWEBMAT. Expert Syst Appl 40(8):2914–2940. https://doi.org/10.1016/j.eswa.2012.12.008

31. Graf S, Viola SR, Leo T (2007) In-depth analysis of the felder-silverman learning style dimensions. J Res Technol Educ 40(1):79–93. https://doi.org/10.1080/15391523.2007.10782498

32. Brusilovsky P (2004) "KnowledgeTree : a distributed architecture for adaptive e-learning". ACM, pp 104–113, https://doi.org/10.1145/1013367.1013386

33. Sung HY, Hwang GJ, Hung CM, Huang IW (2012) "Effect of learning styles on students' motivation and learning achievement in digital game-based learning". 2012 IIAI Int Conf Adv Appl Informatics, pp 258–262, https://doi.org/10.1109/iiai-aai.2012.59

34. Özyurt Ö, Özyurt H (2015) "Computers in human behavior learning style based individualized adaptive e-learning environments : content analysis of the articles published from 2005 to 2014". 52:349–358, https://doi.org/10.1016/j.chb.2015.06.020

35. Popescu E (2010) Adaptation provisioning with respect to learning styles in a Web-based educational system: an experimental study. J Comput Assist Learn 26(4):243–257. https://doi.org/10.1111/j.1365-2729.2010.00364.x

36. Yang T, Hwang G, Yang SJ (2013) "Development of an adaptive learning system with multiple perspectives based on students' learning styles and cognitive styles". 16(2):185–200

37. Gilbert JE, Han CY (1999) "Arthur: adapting instruction to accommodate learning style", p 7

38. Wolf C (2003) "iWeaver : towards ' learning style ' -based e-learning in computer science education". *Fifth Australas Comput Educ Conf (ACE 2003), 2003 Aust Comput Soc 2003 CRPIT ISBN 0-909925-98-4*, 20:273–279

39. Carver CA, Howard RA, Lavelle E (1996) Enhancing student learning by incorporating learning styles into adaptive hypermedia. Proc Ed-Media 96:118–123

40. Triantafillou E, Pomportsis A, Georgiadou E (2002) "AESCS: adaptive educational system base on cognitive styles". In *Proceedings of the AH2002 Workshop*, Malaga, Spain, pp 10–20

41. Cristea A (2004) "Authoring of learning styles in adaptive hypermedia : problems and solutions". Learning May:114–123, https://doi.org/10.1145/1013367.1013387

42. Stern MK, Woolf BP (2000) "Adaptive content in an online lecture system". *International conference on adaptive hypermedia and adaptive web-based systems*, pp 227–238

43. Grigoriadou M, Papanikolaou K, Kornilakis H, Magoulas G, "2 INSPIRE ' s adaptive functionality". Artif Intell

44. Carro RM, Pulido E, Rodríguez P (2015) "TANGOW : task based adaptive learNer guidance on the WWW," pp 1–10

45. Lee G, Lee J, Kim DC (2012) "International conference on future computer supported education, August 22–23, 2012, Fraser place central—seouladaptive learning system applied bruner' EIS theory," *IERI Procedia*, vol. 2, pp 794–801, doi:http://dx.doi.org/10.1016/j.ieri.2012.06.173

Chapter 4
Learning Style with Cognitive Approach

4.1 Learning Style with Learning Theory

4.1.1 Learning Style

The learning styles are domain of cognitive, affective, and psychological character-istics of the learners. It is all about how a learner understood the study materials, perceived the contents, interacted with learning objects, and responded to the tests and applications in the given learning environment with different stimuli (Keefe 1979).

4.1.2 Learning Styles Theories

The first step of integrating process of learning style is to design a framework and select the most appropriate LS theory. In the last 30 years, more than 70 LS theories were formulated by different scholars which are discussed by Coffield et al. [1]. A few theories are common in many ways. But most of them are critical in view of validity and reliability. However, some theories have been derived and implemented in adaptive e-learning.

Learning style classifiers are introduced in learning style-based AES as shown in Fig. 4.1 of Sect. 4.1.3. Figure 4.2 shows the most frequently used learning style classifiers based on previous work.

© The Author(s), under exclusive license to Springer Nature Singapore Pte Ltd. 2021   39
S. Sweta, *Modern Approach to Educational Data Mining and Its Applications*,
SpringerBriefs in Computational Intelligence,
https://doi.org/10.1007/978-981-33-4681-9_4

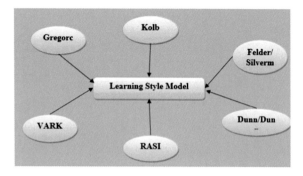

Fig. 4.1   Six prominent learning style models

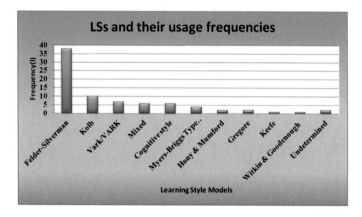

Fig. 4.2   Learning styles and their usage frequencies

4.1.3   Six Prominent Learning Style Models

See Figs. 4.1 and 4.2.

4.1.4   Discussion on Existing Personalized Adaptive E-Learning System

Bergasa-Suso et al. [2]: In this research study, scholars conducted a study taking ILS (Training) questionnaire with 67 learners and i-lessions with seven learners. They used browser-based technology system with specified rules and applied data-driven approach to process learning dimensions.

Limitations: It is related to Internet access and providing recommendations of pages similar to the browsing history or other relevant pages. It lacks adaptive e-learning and does not fit for any serious and structured courses or learning materials.

Cha et al. [3]: The scholars applied the concepts of decision trees (DT) and hidden Markov models (HMM) for identifying learning styles as per FSLSM using data-driven approach.

Limitations: Due to the restrictions of using specific data, the study is limited only to the formation of learners' learning models.

Garcia et al. [4]: This study was carried on the basis of data-driven approach. The researchers used SAVER system as an AI web-based assessment tool and applied Bayesian networks as technological tool to detect learning styles of 27 system engineers based on their observed behaviors during the online course.

Limitations: Only three dimensions of FSLSM have been considered in this study. The fourth dimension visual/verbal was not taken into account which is considered as an important dimension in latest studies. The outcomes showed that the application of BN holds better only in sensing-intuitive dimension. On the contrary, learners must have prior knowledge of computer and internet and involved to interact with one another for detecting other dimensions.

Garcia et al. [5]: This study is based on a data-driven approach in which Bayesian networks was used and feedbacks from 42 system engineers were taken for experimentations and deliberations. In this study, a concept of e-teacher was introduced. It is like an intelligent agent which ensures personalized support to e-learning learners. The e-teacher monitors learner's characteristics while doing e-learning courses, builds learner's profile, and suggests personalized courses. It comprises learners' learning style and their performance in terms of number of visits, completed exercises or application, topics selected, session studied, examination results declared, etc. The scholars examined the behaviors of learners in online course in the SAVER system and conducted two experiments to prove the effectiveness of Bayesian networks for identifying learning styles based on learner's behavior.

Limitations: The concept of e-teacher does not provide a holistic approach. It only covers how to assist and suggests learners the personalized course of action. It does not conceptualize and implement the FSLSM's dimensions considerably. It does not cater about providing adaptive learning path and materials.

Graf et al. [6]: This study was based on literature-based approach using FSLSM. The research scholars observed the behavior of 127 learners and evaluated the LSs through an object-oriented course in LMS on Moodle. Threshold values were predefined for different activities on LMS for capturing behavioral patterns.

Limitations: This study shows moderate precision in terms of sequential /global dimension of FSLSM. In real-life scenario, adaptivity is of dynamic nature. Dynamic adaptivity was not considered while detecting LSs and providing adaptive learning materials to the learners.

Ozpolat and Akbar [7]: The scholars proposed an automated learning model in which diagnosis and classification of learning styles were done by the methods of NB-Tree and binary relevance classifier.

Limitations: Only information-related data objects selected by the learners were processed. Other influential factors for adaptive e-learning were ignored. Results related to input dimensions were not very encouraging. The scholars advised to explore the usage of fuzzy reasoning engines and stressed upon using the hybrid model in a more precise way.

Chang et al. [8]: This study was based on data-driven approach in which LSs were identified and classified as per proposed model. The significance of adaptive e-learning was emphasized in which learning materials/strategies/paths provided to the learners based on their LS.

Limitations: Any specific learning style model was not considered for this study. Adaptivity in e-learning environment was not focused as desired. Dynamic adaptivity is also not applied in the proposed mechanism.

Simsek et al. [9]: This study was based on the literature-driven approach in which prediction of learning style was carried out by automated student's modeling and learner interface interactions. In this model, first the behavior patterns and their affixed threshold values were calculated. Subsequently, explicit rules were applied to find out the learning styles from the measurable clues of learners' behavioral.

Limitations: Only one dimension of FSLSM was used in the study. The study does not fit in category of holistic approach based on a select group of students of mathematics who showed specific characteristics. The discussion and implementation of dynamic adaptation were not considered.

Darwesh, Rashad, and Hamada [10]: In this study, data-driven approach was used and the scholars proposed a LMS. The model detects the learning by analyzing the user behavior through the interactions with web pages (like www.tagmel.com) using social bookmarking software.

Limitations: The study was improved by applying learning vector quantization. The methodology used as ILS filled questionnaire in social bookmarking sites could not ensure detection of learning styles precisely and could not provide adaptive e-learning.

Darwesh, Rashad, and Hamada [11]: This study was extension of earlier work [10]. In this study, web pages tagging was done by using social bookmarking and learning vector quantization. It was data-driven and LMS-independent approach.

Limitations: It has neither provided any holistic approach nor implemented any learning style model. It does not cover adaptive e-learning aspects.

Montazar and Ghorbani [12]: In this study, the authors formulated an evolutionary fuzzy clustering (EFC) mechanism with added advantages of genetic algorithm (GA).

Limitations: People with similar learning styles may form a cluster, but adaptive e-learning a different aspect which was not touched upon by them.

Deborah et al. [13]: This was a data-driven study using FSLSM dimensions and applying fuzzy logics. The researchers had taken base model proposed by Sanders and others and evaluated the students' classification for C programming language course of Computer Science and Engineering Students, Anna University, Chennai. In this study, bell-shaped membership function was used which showed better classification for "unknown."

Limitations: Fuzzy rules were applied to handle uncertainties only to very limited purposes. The study did not implement their complete mechanism like fuzzification, applications of fuzzy rules, and defuzzification to get the result in comparable form of studies. Adaptive e-learning approach was also not well taken.

Dung and Florea [14] Literature-based approach was used in this paper as devised by Graf [15]. The learning style preferences were detected by evaluating no. of accesses to the LO and time spent there by a learner.

Limitations: This study was extension of the work done by Graf and others in which two variables are taken into consideration for detection of learning styles. It only discussed detection of learning style and did not provide any idea about adaptive e-learning.

Montazer and Saberi [16]: This study was based on data-driven approach in which a framework was proposed by taking advantages of fuzzy learner modules and optimized fuzzy item response theory (OFIRT) teaching methodology. This study was a combination of ILS, LMS Logs, and Bayesian networks.

Limitations: The study was limited to monitoring, identifying, and evaluating the learners. It lacks in providing Personalized Adaptivity in e-Learning System.

Table 4.1 list includes several researches that focused on diagnosing learners' learning style from 2005 to 2012 given in next page. Numerous methodologies, models, frameworks, DM techniques, theoretical and practical approaches, and various objectives of finding distinct learning styles had been studied, extracted the facts, built conceptual learning, and implemented though proposed framework with some identified methodologies with proper classifications. Many inferences had been implied in the study, but the most prominent techniques used to understand the patterns and deciphering the individual's learning styles in e-learning are statistical and mathematical analysis, FCMs, fuzzy logics, neural network; feed-forward neural networks; decision tree; Bayesian networks; genetic algorithm; rough sets; and association rule.

Research orientations can be easily understood from the data patterns and analysis that finding out learner's behavior has been trusted upon capturing the learner's interactions and their profile data in learning management system (LMS). The flexibility and robustness provided by LMS made it a successful tool to develop e-learning system. They possess the log files containing the stored interactions data and their tracking history of the learner. These data are utilized to filter the learner's behavioral traits in e-learning system. The research was carried out on classifying student's learning style as per direction of Felder–Silverman model [18], and four classifiers or dimensions were required to map the learner's characteristics accordingly.

These studies are targeted to take holistic views all about the learning objects, course contents, academic achievement, knowledge levels and key learning factors, learning input–output dimensions, usability or preferability to determine the satisfaction levels by automatic detection of learning styles. Some of these researches had proposed suitable models and developed structural frameworks, whereas some were carried out experiments with the AEHs. These experimental studies were of utmost important to determine the effect of such actions and validated the proposed models or frameworks [19, 20] or explored inferences for further fine tuning.

Table 4.1 Comparative analysis of Literature Review

No	Research studies	Rules or techniques	Main points	Approach	Assessment methodology	Precision %
1.	Bergasa-Suso et al. [2]	Browser-based System with Rules	Processing Dimension	Data-Driven	ILS Training model: 67 students & in iLessons: 7 students	71% in Processing
2.	Garcia et al. [4]	BN	Detection only	Data-Driven	AI—SAVER: 27 system engineers	58% in Processing 77% in Perception 63% in Understanding
3.	Garcia et al. [5]	BN	Detection + Suggestions	Data-Driven	AI—SAVER with e-Teacher: 42 systems engineers	83% feedback received was positive
4.	Graf et al. [6]	Simple rules on Matching Hints	LMS Independent Better than data-driven Approach	Literature-Based	Info. Sys. & Comp. Sci.—Austria Univ.—Object-Oriented Modelling—Moodle LMS—127 students	77.33% in Input 79.33% in Processing 76.67% in Perception 73.33% in Understanding
5.	Ozpolat and Akbar [7]	NB-Tree classification with Binary Relevance Classifier	Detection + Suggestion data objects selected by the user in LMS independent	Data-Driven	10 graduate students (Training) 30 graduate students (Testing)—PoSTech	53.3% in Input 70% in Processing 73.3% in Perception and Understanding
6.	Chang et al. [8]	Enhanced k-NN Clustering with GA	k-NN—Pre-Contrast & Post-Comparison Decrease in no. of features	Data-Driven	SCORM- Javabased LMS—Windows XP IRIS data set by UCI 117 students -	Increasing Accuracy

(continued)

Table 4.1 (continued)

No	Research studies	Rules or techniques	Main points	Approach	Assessment methodology	Precision %
7.	Simsek et al. [9]	Simple rules based on mapping	6 features considered of processing dimensions	Literature-Based	27 students—Comp. Educ.—Derivatives—Moodle LMS	79.63% in Processing
8.	Darwesh, Rashad, and Hamada [10]		ILS filled using social bookmarking site	Data-Driven	25 and 15 students in two sets www.tagme1.com	For No. of learners = 25, Recognition = 72% For No. of learners = 15, Recognition = 86.66%
9.,	Darwesh, Rashad, and Hamada [11]	Web pages tagging on social bookmarking and Learning Vector Quantization	Auto collected data: LMS independent	Data-Driven	By changes in Rate of Learning, No. of hidden neurons & values of epochs	For Learning Rate = 0.01, Number of Hidden Neurons = 40 and Epochs = 150, Recognition Rate = 93.33% for 15 learners
10.	Montazer and Ghorbani [12]	Evolutionary Fuzzy Clustering (EFC) using Genetic Algorithm	People with similar LSs in a cluster, High computational and memory usage costs, used Particle Swarm Optimization technique	Data-Driven	Fundamentals of Computer Networks: 98 undergraduate students	EFC more accurate than grouping based on behavior in log files of LMS
11.	Deborah et al. [13]	Fuzzy Logic	Bell-shaped Membership function Good for "Unknown" variables	Data-Driven	Comp. Sci. & Engr.—Anna Univ.—C-language	-NA-

(continued)

Table 4.1 (continued)

No	Research studies	Rules or techniques	Main points	Approach	Assessment methodology	Precision %
12.	Dung and Florea [14]	Simple rules on Mapping things	LMS Independent Parameters. Only taken No. of visits and Time spent	Literature-Based	Comp. Sci.—Politechnica Univ., Bucharest—AI course—44 UG students Web-based LMS POLCA	70.15% in Input 72.73% in Processing 70.15% in Perception 65.91% in Understanding
13.	Montazer and Saberi [16]	ILS + LMS logs + BN	Improved accuracy and Decreased uncertainty	Data-Driven	Study conducted in 3 phases: 40 M.Sc. students on four different courses	-NA-
14.	Jyothi et al. [17]	Recommender System on ILS and Clustering basis	150 users Prior knowledge of learner required to detect LS by ILS	Data-Driven	C-DAC: 105 students Hyderabad R&D labs—R&D + courses like Embedded Systems, System S/W and Adv. Business Computing	Good accuracy data sets less than 150

The studies related to constructing and applying learning styles and adaptive e-learning system had many positive outcomes were discussed [4, 21] and their applications as providing personalizing learning materials and learning contents were also discussed and developing educational games were explored in [22].

Some researchers who supported the importance of the learning factors in learning processes are discussed as under noted: -

- Essalmi et al. in [23], "Affective states, mindset and learning styles tactics to provide personalization in e-learning system have a significant effect on student learning in process and performance."
- Graf et al. in [15], "Personalization based on detection of LSs ensures less timing in study, enables average learner to perform better and grab higher grades, more learning satisfaction."
- Sfenrianto et al. [24] touched upon "identifying triple-factor based on learning behaviors of the learners in context of dynamic personalization in e-learning process."

4.2 Comparative Analysis of Learner Modeling Techniques

Selection of an appropriate technique for the required uses is paramount, though it is considered as a challenging task. Each and every applications of adaptive e-learning system is important in terms of the required level of adaptation, valid rules for adaptation, availability of training data [25]. It presents some instructions to how to choose the technique in the light of the proposed framework and set objectives. The dimensions considered for comparison are listed below:

- Analytical criticality relates to the time spent in offline processing.
- Dynamic modeling refers the ability of the methodology to change the learner model while the learning process is in full swing.
- Labeled or unlabeled data refer to the type of data used in selection.
- Size of training data refers to their validity in adaptation process.
- Uncertainty refers to how to handle uncertainties and what are remedial actions required.
- Noisy data refer to how they affect the model and its smooth functioning.

Table 4.2 compares techniques used in different learners' modeling. They recommend which technique to be selected according to what adaptive application is most suited [26, 27].

Table 4.2 Characteristics of different techniques of learners modeling

Technique	Complexity	Dynamic modeling Yes or No	Labeled/ Unlabeled	Size of training data	Uncertainty Yes or No	Noisy data Yes or No
Fuzzy rules	Medium	Y	NA	NA	Y	Y
ANN	High	Y	All	High	Y	Y
Fuzzy clustering	High or Med	N	All	Med or High	Y	Y
Neuro-Fuzzy	High	Y	Lbld.	Med or High	Y	Y

References

1. Coffield F, Moseley D, Hall E, Ecclestone K, Coffield F (2004) "Learning styles and pedagogy in post-16 learning: a systematic and critical review", Accessed 03 Oct 2020. [Online]. Available: http://evidence.thinkportal.org/handle/123456789/62
2. Bergasa-Suso J, D. S.-I. T. on, and undefined 2005, "Intelligent browser-based systems to assist Internet users," *ieeexplore.ieee.org*, Accessed 03 Oct 2020. https://ieeexplore.ieee.org/abstract/document/1532366/
3. Truong HM (2015) "Integrating learning styles and adaptive e-learning system: current developments, problems and opportunities", Comput Human Behav. https://doi.org/10.1016/j.chb.2015.02.014
4. García P, Amandi A, Schiaffino S, Campo M (2005) "Using bayesian networks to detect students ' learning styles in a web-based education system," 11, 29–30
5. Schiaffino S, Garcia P, Amandi A (2008) eTeacher: providing personalized assistance to e-learning students. Comput Educ 51(4):1744–1754. https://doi.org/10.1016/j.compedu.2008.05.008
6. Graf S (2008) "Identifying learning styles in learning management systems by using indications from students' behaviour Kinshuk Tzu-Chien Liu", pp 482–486, https://doi.org/10.1109/icalt.2008.84
7. Özpolat E, Akar GB (2009) Automatic detection of learning styles for an e-learning system. Comput Educ 53(2):355–367
8. Chang YC, Kao WY, Chu CP, Chiu CH (2009) A learning style classification mechanism for e-learning. Comput Educ 53(2):273–285. https://doi.org/10.1016/j.compedu.2009.02.008
9. Ömer Ş, Atman N, Murat M, Arikan YD (2010) "Diagnosis of learning styles based on active / reflective dimension of felder and silverman' s learning style", pp 544–555
10. Hamada AK, Rashad MZ, Darwesh MG (2011) Behavior analysis in a learning environment to identify the suitable learning style. Int J Comput Sci Inf Technol 3(2):48–59. https://doi.org/10.5121/ijcsit.2011.3204
11. Darwesh MG, Rashad MZ, Hamada AK (2011) From learning style of webpage content to learner' s learning style. Int J Comput Sci Inf Technol 3(6):195–205. https://doi.org/10.5121/ijcsit.2011.3615
12. Ghorbani F (2011) "Learners grouping in e-learning environment using evolutionary fuzzy clustering approach". Accessed 03 Oct 2020
13. Jegatha Deborah L, Baskaran R, Kannan A (2014) Learning styles assessment and theoretical origin in an E-learning scenario: a survey. Artif Intell Rev 42(4):801–819. https://doi.org/10.1007/s10462-012-9344-0
14. Dung PQ, Florea AM (2012) An approach for detecting learning styles in learning management systems based on learners' behaviours. Int Conf Educ Manag Innov 30:171–177

15. Graf S, Viola S, Kinshuk (2007) "Automatic student modelling for detecting learning style preferences in learning management systems", IADIS Int Conf Cogn Explor Learn Digit Age 1988:172–179, http://sgraf.athabascau.ca/publications/graf_viola_kinshuk_CELDA07.pdf
16. Saberi N, Montazer G (2012) "A new approach for learners' modeling in e-learning environmentusing LMSlogsAnalysis", 2012 Third Int Conf E-Learning E-Teaching, pp 25–33
17. Jyothi N, Bhan K, Mothukuri U, Jain S, Jain D (2012) "A recommender system assisting instructor in building learning path for personalized learning system", Proc.—2012 IEEE 4th Int Conf Technol Educ T4E 2012, pp 228–230, https://doi.org/10.1109/t4e.2012.51
18. Felder RM (2005) "A study of the reliability and validity of the felder-soloman index of learning styles", Eng Educ 113:77 [Online]. Available: http://www4.ncsu.edu/unity/lockers/users/f/fel der/public/ILSdir/Litzinger_Validation_Study.pdf
19. Sweta S, Lal K (2016) "Learner model for automatic detection of learning style using FCM in adaptive e-learning system", 18(2):18–24, https://doi.org/10.9790/0661-1802041824
20. Özyurt Ö, Özyurt H (2015) "Computers in human behavior learning style based individualized adaptive e-learning environments : content analysis of the articles published from 2005 to 2014", 52:349–358, https://doi.org/10.1016/j.chb.2015.06.020
21. Graf S, Lan CHLCH, Liu T-CLT-C, Kinshuk K (2009) "Investigations about the effects and effectiveness of adaptivity for students with different learning styles", 2009 Ninth IEEE Int Conf Adv Learn Technol, pp 415–419 https://doi.org/10.1109/icalt.2009.135
22. Dascalu M-I, Bodea C-N, Moldoveanu A, Mohora A, Lytras M, de Pablos PO (2015) A recommender agent based on learning styles for better virtual collaborative learning experiences. Comput Human Behav 45:243–253. https://doi.org/10.1016/j.chb.2014.12.027
23. Essalmi F, Ben Ayed LJ, Jemni M, Graf S, Kinshuk (2015) "Generalized metrics for the analysis of E-learning personalization strategies", Comput Human Behav 48:310–322, https://doi.org/10.1016/j.chb.2014.12.050
24. Sfenrianto, Suhartanto H, Hasibuan Za (2012) "A dynamic personalization in e-learning process based on triple-factor architecture". Comput Technol Inf Manag (ICCM), 2012 8th Int Conf 1:69–75
25. Ahmad A, Basir O, Hassanein K (2004) "Adaptive user interfaces for intelligent e-learning : issues and trends". Fourth Int Conf Electron Bus, pp 925–934
26. Bralić A, Ćukušić M, Jadrić M (2015) "Comparing MOOCs in m-learning and e-learning settings," no. May, pp. 25–29
27. Project MT, Report SI (2013) "System for mooc"

Chapter 5
Framework for Adaptive E-Learning System

5.1 Introduction

Latest technology has opened floodgate of opportunities to establish new educational system which can provide Personalized Adaptive e-Learning in recent years. Now, learners can learn their desired courses anywhere and anytime. It is very critical to facilitate personalized learning material to every learner as per their own preferences. The learners may come from different backgrounds and having different profile data and so having different behavioral characteristics which can be exhibited during the e-learning courses. To address these issues, many scholars devised frameworks to cater the needs of learners as well as to fulfill the objective of academician and educational institutions to make them learn adaptively. Machine learning-based algorithm was also required to acquire, diagnose, and provide personalized adaptively in addition to the frameworks.

The objective of a framework discussed here is to detect learning style, diagnose preferences, provide personalization, and provide complete dynamic adaptation by using machine learning techniques. It is imperative to select heterogeneous group of learners having different knowledge, skill, attitudes, preferences, goals, and plans. They may exhibit different interactive behaviors and different usages data. These parameters help to find patterns, and they can be analyzed to detect the distinct learning style and its preferences. A learning system can be developed in which adaptation changed as per responses and interaction in between learners and system during entire learning processes. Many learning objects can be identified, put in place appropriately and provided to find out different learners' preferences for one or more learning objects. In the system, it can be ensured that the contents are being provided according to the learners' need and same content could not be provided to all. Such system can be designed to reach many learners. It will be very cost effective also. Therefore, the learning process may be better and encouraging because of the uses of dynamic adaptations.

S. Sweta, *Modern Approach to Educational Data Mining and Its Applications*,
SpringerBriefs in Computational Intelligence,
https://doi.org/10.1007/978-981-33-4681-9_5

5.2   An Adaptive Framework

With the help of adaptive e-learning methods, an adaptive framework can be enabled to provide best-suited learning materials based on analyzed data and patterns.

The architecture of the framework can be devised as given below: (Fig. 5.1)

Personalized Adaptive Learner Model (PALM) captures learners' profile and behavioral data during learning process of a course on basis of observed learning behavior, diagnoses learning style preferences and provides personalized adaptive learning materials to each individual's learners according to their identified preferences. It improves the better communication and interaction between learner and the system and enhances the learning process for making efficient learning system.

As an author of this book devised a framework consists of a comprehensive learner model comprising four modules.

They are:

1. Learner Profile and Interface Module (LPIM)
2. Behavior Monitoring Module (BMM)
3. Learning Style Diagnostic Module (LSDM)
4. Personalized Adaptive Module (PAM).

This model PALM is based on Felder–Silverman LS Model. For testing this model, there are two approaches applied in this adaptive e-learning system. One is teacher

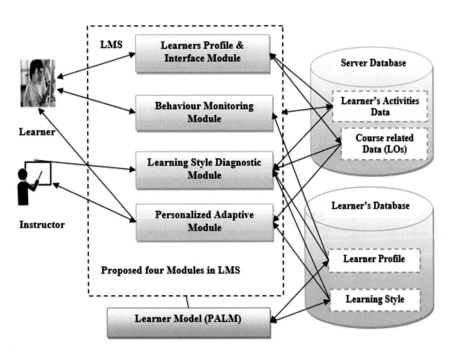

Fig. 5.1 A framework

centric and another one is learner centric. In first approach, learning material is being provided by the system or the instructor in a non-adaptive manner, whereas in second approach it is by the system in a dynamic and adaptive way according to identified learning style preferences of the learners. The learner model is dealt by considering of all the characteristics of learners, their interactions, and behavioral aspects during the course to meet the objective of learning effectively. Modeling learner behavior is one of the most important task before providing personalization [1]. Learners preferences are also an important aspect which is primarily provided by the learners themselves [2]. Learners-related data are of two types one is provided by learner that is stored in his/her profile data it is also known as static data and another one is dynamic data captured by the system during learning course materials i.e. learner's behavioral pattern. These data are analyzed by the learner model to meet the objectives of the study. Customization is based on individual learning style, selection of learning material and interaction log created during learning. Adaptive system provides optimization of various input inferences and apt learning contents to the learners. The model identifies complex data, correlates them to do fruitful analysis, and recommends different learning styles. However, crisp value does not make good analysis of behavioral patterns of human. Moreover, fuzzification helps to tackle such complex data just like human brain as it works in a cognitive way. The set of data structure is critical for storing learner's behavior data, their preferences, and demographic data relevant to their performance and adaptive learning.

5.2.1 *Learner Profile and Interface Module (LPIM)*

The Learner Profile and Interface Module (LPIM) gives the method as interface for collecting all relevant data related to learner's characteristic and provides highly interactive user-friendly interface. Learners do registration by giving input data in form of a profile data set in learner's profile module, and these data are stored in learner database shown in Figs. 5.5 and 5.6. When learner gets login and put user ID and Password, system asks to change the password. Then learner will be ready to interact with the system after selecting course with agreed terms and conditions shown in Figs. 5.7 and 5.8. In fact, online forms are designed to capture all the demographic data for better understanding the learner's characteristics [3]. These forms are shown in Figs. 5.2, 5.3, 5.4, 5.5, 5.6, 5.7, 5.8, 5.9 and 5.10. The aggregated data are stored in learner database and called as learner profile data. However, these data are static in nature. Learner's dynamic data are dealt with in Behavioral Monitoring Module during sessions.

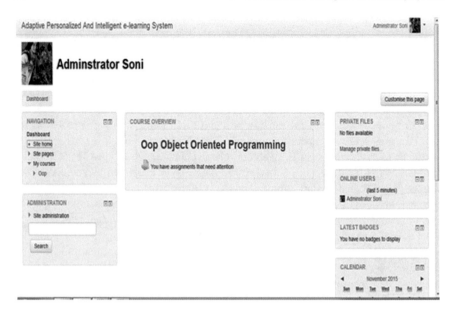

Fig. 5.2   Administrative interface view of APIE LMS

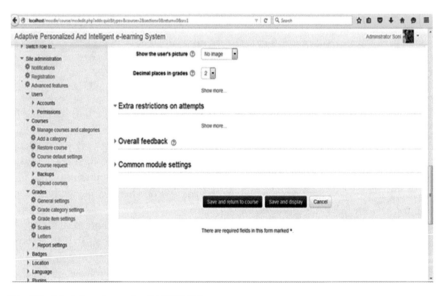

Fig. 5.3   Administrative tools of APIE LMS

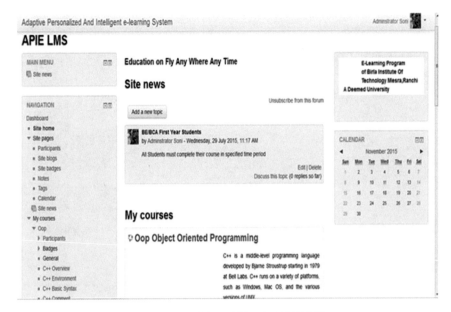

Fig. 5.4 Teacher interface for new course/blog addition in APIE LMS

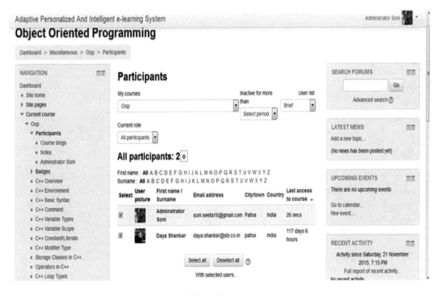

Fig. 5.5 Participants interface view in APIE LMS

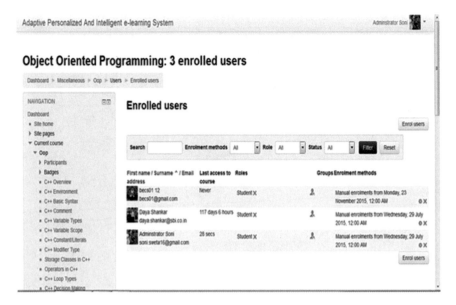

Fig. 5.6   Enrolled students in APIE LMS

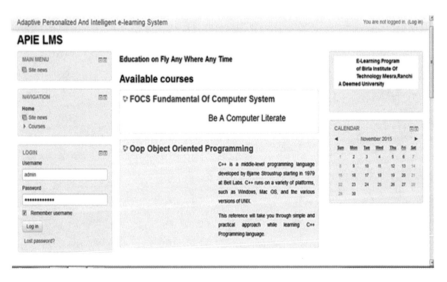

Fig. 5.7   All available courses in APIE LMS

5.2.2   Behavior Monitoring Module-BMM

Data related to learners' characteristics in e-learning system are very much important to diagnose and identify the behavioral patterns of each of the learners in order

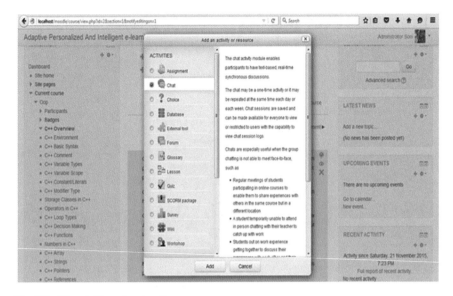

Fig. 5.8 All activities related to Course

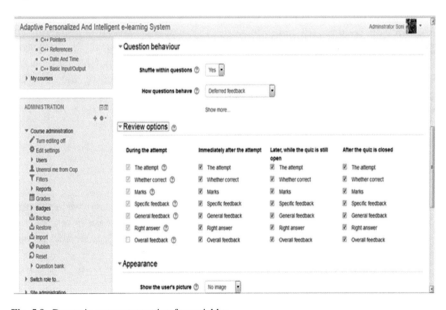

Fig. 5.9 Dynamic parameter setting for variables

Fig. 5.10 Types of resources available in APIE LMS

to achieve the adaptation. The Behavior Monitoring Module (BMM) describes the methods for collecting the all relevant data related to learner's behaviors and characteristics during the e-learning processes. These data are dynamic in nature. The behavior of the learner can be detected during the session automatically by the system. The measurable input data related to learning activities and behaviors shown in Fig. 5.9 can be analyzed and interpreted on the basis of output log file data. The valuable information and deciphered patterns can be obtained, and final findings can be summarized.

5.2.3   Learning Style Diagnostic Module-LSDM

When learners interact with the system, all the interactive and activity data are automatically captured and stored in the databases, i.e., server database in terms of log file and learner's database in the system. Let during interaction learner interact with different learning objects of set L {L1, L2.... Ln}.

However, learner behavioral data which are stored in separate database server gives the specific patterns. These patterns can be analyzed by the Learning Style Diagnostic Module. The overview of idea of identification of learning style preference is given in Fig. 5.11. This module dedicated to LS adaptive parameters which allow adapting the learning materials/objects corresponding to the result of the learner's model.

The fuzzy cognitive map techniques can be applied to find the casual relationship among behavioral data as concepts. The patterns and expert knowledge about the concept value and its interconnected weight helps to design the fuzzy cognitive

Fig. 5.11 Learning style
preference identification

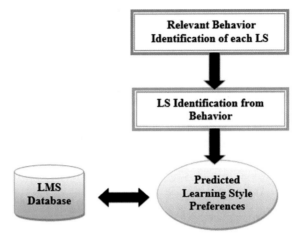

map model. The FCM is also mapped the learning style to learning object for each dimension shown in Chap. 6. As the model is based on FSLSM, it analyses only the relevant variables which directly or indirectly co-related with their dimensions. Learner's preference is an important aspect which can be captured in this module to predict learner's behaviors and which is very much important for making appropriate adaptation in PALM. Thus, human learning is considered as combination of cognitive, affective, and psychological behaviors which perceive learning in an interactive way and responses to the learning environment [4].

5.2.4 Personalized Adaptive Module-PAM

This module provides personalized course contents based on chosen three influential factors in e-learning. The module basically introduces two cognitive factors motivation and knowledge ability with learning style and final learner's learning preferences can be obtained. Qualitative parameters like motivation and knowledge can be measured for better understanding of learner preferences [5]. In this module, two aspects are dealt with—one is personalization or another one is adaptation (automatic customization). In personalization, system is enabled with all about the learners' characteristics, filtered suitable contents, and mapped appearances or behaviors according to their changed preferences. Whereas in adaptation, system is devised with a model for the learner in which automated learning system provided suitable learning materials according to their needs and adjust intermittent situations if arising in between. As it is a dynamic way of adaptation.

5.3　Workflow of Adaptive System Components

- Learners are the users located at distant places and who opted for the e-learning course. They interact with the learning objects linked system applications or instructors during the learning process.
- Instructor is the faculty or the system that is designed to provide learning materials and define processes.
- The learner model is combination of modules providing adaptation to each and every individual as per the learning style preferences.
- Server DB is a database of learning material/activities which capture log files.
- Learner model is aggregation of the learners' individuality in the system based on their traits and shows different learning style preferences (Fig. 5.12).

The adaptive e-learning system framework is empowered to provide personalized learning experiences adequately and learning contents efficiently. It processes the problems as similar as human brain. The way human processes information is called cognitive style (CS) [6]. They acquire information, recognize them discretely, and co-relate them with proper representations to get the knowledge and their usage. Learning is a dual process, namely reception and system generated process or stimuli. Reception can be realized through various senses from external sources, and another process is internal activities like mood, motivation, emotions, sentiments, recalling things, learning by heart, influenced by self or others, self-realization, reflection, and self-actualization. Learners' preferred way of reception and processing is called learning style (LS) [7]. Several learning models and theories are proposed by combining different influential parameters of learning styles. Most of the researchers classified the learners into similar groups. One of the most prominent LS models is Felder–Silverman learning model (FSLM) which defines the characteristics and can be easily translated into our framework. Table 6.1 of Chap. 6 identifies the different characteristics spelt out in the FSLSM [8, 9].

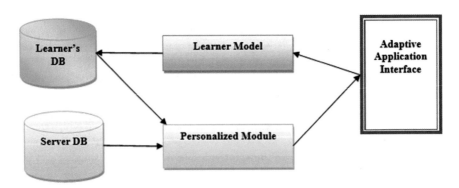

Fig. 5.12　Workflow of an adaptive system

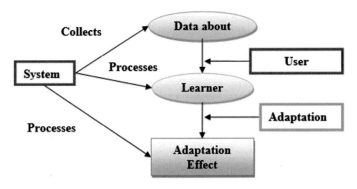

Fig. 5.13 Data flow in adaptive system

5.4 Adaptable Characteristics for Learner Model with System Process

Adaptation decisions do not depend upon only to the known factors, but they are equally dependent upon the various characteristics exhibited by the users known or unknown during interaction with the system [10]. The model is designed to auto-update the contents during the course of interaction between the learner and the system in terms of dynamically changing the concept values and strengthening the entire process. In Chaps. 2 and 3, many related studies are reviewed and the existing techniques in construction of learner's models in the e-learning environment are explained. The data flow of our adaptive system process is shown in Fig. 5.13.

5.5 Factors Affecting on Personalization

There are many factors affecting personalization in e-learning system. Learner profile data stored in learner database helps to make a classification among learners based on these data. When learners interact with the system, log data are generated, log files are stored in server database, and valuable information is processed to get learners' Learning Style Diagnostic Module. The direct feedback from the learners and log data is generated during interaction with the peer groups that is stored in database helps in fine tuning the course material and detect communication aspects. System takes care of these aspects before generating personalization to the learners.

5.6 Factors Affecting Adaptation

In personalization, explicit information is analyzed but in adaptation implicit information about the background knowledge of the learner are also analyzed. In adaptation, background knowledge of learner is required which enables system to provide on the move learning materials. In adaptive system, learners are aware about time and accuracy of the result. Here, variation in input data is of much important that can be generated during the interaction of the system with learners to make useful information. Every effort is made to make the learner's friendly model in which learning centric approach is applied considering the fact that learners differ in knowledge, skill, aptitude, and preferences.

References

1. Huang EY, Lin SW, Huang TK (2012) What type of learning style leads to online participation in the mixed-mode e-learning environment? A study of software usage instruction. Comput Educ 58(1):338–349. https://doi.org/10.1016/j.compedu.2011.08.003
2. Kobsa A (2007) "Generic user modeling systems". Adapt Web (LNCS 4321), pp 136–154, https://doi.org/10.1007/978-3-540-72079-9_4
3. Sweta S (2015), "Adaptive and Personalized Intelligent Learning Interface (APIE-LMS) In e-learning System," vol. 10, no. 21, pp. 42488–42492
4. Villaverde JEÃ, Godoy DÃ, Amandi AÃ (2006) "Learning styles' recognition in e-learning environments with feed-forward neural networks," pp 197–206
5. Marković S, Jovanović N (2012) Learning style as a factor which affects the quality of e-learning. Artif Intell Rev 38(4):303–312. https://doi.org/10.1007/s10462-011-9253-7
6. Yang T, Hwang G, Yang SJ (2013) "Development of an adaptive learning system with multiple perspectives based on students' learning styles and cognitive styles," 16(2):185–200
7. De Palo V et al (2012) "How cognitive styles affect the e-learning process," *Proc 12th IEEE Int Conf Adv Learn Technol ICALT 2012*, pp 359–363, https://doi.org/10.1109/icalt.2012.79
8. Dominic M, Xavier BA, Francis S (2015) "A framework to formulate adaptivity for adaptive e-learning system using user response theory," I.J Mod Educ Comput Sci 1(January):23–30, https://doi.org/10.5815/ijmecs.2015.01.04
9. Reisman S (2014) "The future of online instruction, part 1," Computer (Long Beach Calif) 47(4):92–93, https://doi.org/10.1109/mc.2014.106
10. Millán PB (2007) "User models for adaptive hypermedia and adaptive eduxational systems," Adapt Web, pp 3–53

Chapter 6
Personalization Based on Learning Preference

6.1 Introduction

This chapter describes the most important components of the framework, i.e. the learner model. Learner model is named as Personalized Adaptive Learner Model (PALM) which identifies learners' learning style during the course and provides personalization according to their needs. This model always updates the learner's data as per the identified style and learning process. Here, student model or user model or learner model can be understood as interchangeable, i.e., dynamic. This chapter discusses application of adaptive learning as well as implementation of two core concepts for resolving problems one is personalization and another one is adaptation. Experts who have knowledge and experience on the operation and behavior of the system are involved in the determination of concepts, ideation of interconnections, and assignment of casual fuzzy weights to the interconnections.

6.2 Fuzzy Cognitive Maps

Fuzzy cognitive model (FCM) is one of the most interesting soft computing techniques for modeling learner model. It is a combination of two important theories of neural networks and fuzzy logics which are interdependent exclusively [1]. Its structure represents numeral data as well as linguistic entities which explain the characteristics and the components of complexity in a model. The knowledge, skill, and attitude of the learner, behavioral cues, and man–machine interactions are studied, and information can be extracted within the prowess of the designed model. FCMs were proposed by Kosko [2]. He showed the causal relationship among applied concepts which helped to analyze inference patterns thereof. FCM is recognized as a valuable tool for detecting learning style. It effectively tackles the uncertainty and fuzziness of data log files generated during the diagnosis of the learning style [3]. It

© The Author(s), under exclusive license to Springer Nature Singapore Pte Ltd. 2021
S. Sweta, *Modern Approach to Educational Data Mining and Its Applications*,
SpringerBriefs in Computational Intelligence,
https://doi.org/10.1007/978-981-33-4681-9_6

possesses several desirable components like easy to design structured knowledge and provides adequate inferences while comparing them along with the expert system or neural networks separately [1].

6.2.1 FCM Outlines and Narration

FCM is originated from the concept of the graph theory. Euler postulated the graph theory in 1736 [1], and afterward, the directed graphs were applied for understanding of the components of the primary data set. Signed directed graphs were used to depict the affirmation of information and cognitive mapping analyzed the graphs in terms of cause–effect relationships among variables. Kosko described cognitive map modeling with two important characteristics: (a) Causal relationships among nodes were fuzzified, and (b) the system can be enabled with the dynamic feedback. The changes across the nodes, their aftereffect at the nodes affecting the initial node and that again initiate the change. The FCM layout is like as recurrent artificial neural network, where concepts are shown by neurons and causal relationships by values of interlinked weights connecting the neurons. The concepts replicate behavioral traits, attributes, features, and senses within the system.

6.2.2 Knowledge Representation with FCM

FCM segregated the entire system into set of concepts and relationships among these concepts. The concept is linked to each node of the FCM. The co-relations among different concepts are shown by the directed arcs [4, 5]. Interconnections among the concepts of FCM denoted the cause and effect relationship. These weighted interconnections have the direction and degree of weight where one concept influences the value of the other concepts and vice-versa. Figure 6.1 shows the graphical representation of a FCM.

FCMs have some remarkable characteristics, like [6]:

1. FCMs capture much information at concept nodes and in between them.
2. FCMs work in dynamic system.
3. FCMs extract and express hidden patterns and relationships.
4. FCMs can be interconnected and can infer valuable inputs.
5. FCMs can be fine-tuned according to the situation.

Fig. 6.1 Fuzzy cognitive map

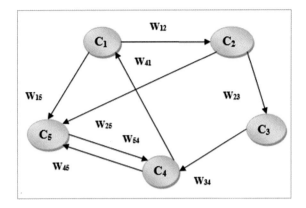

6.2.3 Different Mathematical Representation of Fuzzy Cognitive Maps

6.2.3.1 FCM: Method I

Fuzzy cognitive maps are fuzzified concepts with interconnected degree of strength with feedback mechanism [7]. In Fig. 6.1 given below.

In generalized term, the concept value is represented by concept C_i and concept C_j and interconnected strength value e_{ij} by the weight w_{ij}. The strength value represents degree of causality between two concepts, and values are fuzzified in range $[-1\ 1]$. If $w_{ij} > 0$, it shows positive causality concept value C_i increases, and hence, the C_j. If $w_{ij} < 0$ shows negative causality in which concept value C_i decreases on rise of C_j and vice-vesa.

The concept values are evaluated by applying the equation given below:

$$x_i(t) = f\left(\sum_{\substack{j=1 \\ j \neq 1}}^{n} x_j(t-1)w_{ji}\right) \tag{6.1}$$

where $x_i(t) = C_i$ at time t, $x_j(t-1) = C_j$ at time $t-1$,

w_{ji} = weight of the interconnected C_j and C_i and f = a sigmoid function, $f = \frac{1}{1+e^{-\lambda x}}$.

Other proposed squeezing functions are the $\tanh(x)$, $\tanh(x/2)$, and they provide values in the fuzzy range of $[0, 1]$ or $[-1, 1]$, where the concept values in the range[7].

After a time, values of concepts change, and process recalculates the new concept value as per above equation. The new values for the concepts are evaluated in each simulated step in FCM. It consists of nx_i vector X that accumulates the values of n concepts, the matrix $W = \left[w_{ij}\right]_{1 \leq i, j \leq n}$ which included the values of the causal edge weights and the dimension of the matrices equates to the different concepts n. So, the new state vector X at time t in FCM can be obtained as:

$$X(t) = f\left(W^T X(t-1)\right) \tag{6.2}$$

6.2.4 Structure and Learning from FCM Method I

In the model, the experts who possess diversified knowledge and vast experiences are involved in guiding operations and incorporating behavioral system while determining fuzzy concepts and interconnected weights in FCM [8].

The interconnected weights change in FCM with a first-order learning law as per the correlation or differential Hebbian learning law:

$$W' = -W_{ij} + x_i' x_j' \tag{6.3}$$

So $x_i' x_j' > 0$ if C_i & C_j move in the same direction and $x_i' x_j' < 0$ if C_i & C_j move in the reverse directions.

Therefore, if the values of the concepts are moved in same style, they have strong positive weights in degrees of membership function and vice versa.

6.2.5 FCM: Method II

In the dynamic FCM, concept value of each node changes in which the last value of each concept is instrumental in determining the new value of the concepts. Therefore, the new values of concepts lightly changed after each simulated step.

The derived formula can be applied as given below:

$$x_i(t) = f\left[k_1 \sum_{\substack{j=1 \\ j \neq 1}}^{n} x_j(t-1)w_{ji} + k_2 x_i(t-1)\right]. \tag{6.4}$$

where

$x_i(t)$ C_i at time t,

$x_i(t-1)$ C_i at time $t-1$, $x_j(t-1) = C_j$ at time $t-1$.

W_{ji} the weight between C_j to C_i and f = a threshold function.

k_2 parameter instrumental for getting new concept value based on previous one and.

k_1 parameter influences the interconnected concepts in the structure of the new value of the concept $x_i . k_1$ and k_2 satisfy the equation:

$$x(t) = f\left[k_1\left(W^T X(t-1)\right) + k_2 X(t-1)\right)\right] \tag{6.5}$$

Therefore, Eq. 6.5 evaluates the new state vector X derived from adding the new value concept obtained from the multiplication of the previous concept value at time t, edge matrix W with k_1 factor and a fraction of the past values of concepts with k_2 parameter.

6.2.6 FCM: Method III

It is proposed that a concept can be added on its own past value plus a weight W_{ij} [9]. Now, the equation changes as:

$$x_i(t) = f\left[\sum_{\substack{j=1 \\ j\neq 1}}^{n} x_j(t-1)w_{ji} + W_{ij}k_2 x_i(t-1)\right] \tag{6.6}$$

where

x_i	C_i at time t,
$x_i.(t-1)$	C_j at time $t-1$, $x_j(t-1) = C_j$ at time $t-1$, and.
W_{ji}	weight C_j to C_i, W_{ij} = weight, previous value of concepts applied for new one and f = a threshold functions.

Now, a more compact equation is:

$$x(t) = f\left(W^T X(t-1)\right) \tag{6.7}$$

where W = nonzero diagonal elements.

In the process of the defuzzification of Center of Area (CoA) [10], fuzzy weights are transformed to a numerical weight W_{ji} in the interval $[-1, 1]$ and endorse the overall suggestion of experts. Thus, $w^{initial} = [W_{ji}]$, $i, j = 1, ..., N$, with $= 0$ W_{ii} $= 0$, $i = 1,..., N$, are calculated. Learning procedures are designed to increase the efficiency and robustness of FCMs by regulating the FCM weight matrix [1].

6.3   The Felder–Silverman Learning Style Model

A great deal of similar models had been proposed by Kolb [11], Honey and Mumford, Myers–Briggs, and Felder–Silverman [9]. The FSLSM [12] can be taken into consideration for detecting LS for following reasons [13]:

- The most suitable for educational systems and especially for engineering students,
- *It* describes learning style classification in four dimensions.
- *It* describes cognitive style in more detail while deriving LS.
- *It* illustrates adequate preferences [14], 15.

Learning style (LS) of a learner is a way how a learner collects, processes, and organizes information. According to FSLSM [13], 12, each learner has a preferred mode of learning style measured in four dimensions (active/reflective, sensing/intuitive, visual/verbal, sequential/global) [16].

- ***Definition***: A learning style has 4-tuples. Therefore, $LS = (D1, D2, D3, D4)$ where D_i [-11, $+11$], $i = 1,\ldots,4$ where each D_i is a Felder and Silverman Learning Style Dimension, i.e., D_1:A/R, D_2:S/I, D_3:Vi/Ve, and D_4:S/G.

A Questionnaire: Index of Learning Style (ILS) consists of 44 questions, i.e. 11 from each of the LS dimensions.

6.3.1   Scales of the Dimensions

Felder's model contains 32 learning styles. Each learning style is defined by the responses in a way of quest of five problems [12] (Fig. 6.2):

1. What type of information the learner preferably **perceived?**
   Sensory (external) vision, acoustics, sensations, or **Intuitive** (internal) inferences, insights, etc.
2. Through which sensory channel external information most effectively perceived?
   Input: visual videos, pictures, diagrams, graphs, or **verbal** words, sounds, etc.
3. With which organization of information, the learner most comfortable? **inductive or deductive**

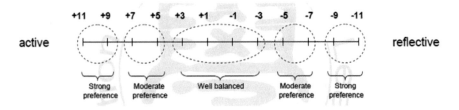

Fig. 6.2   Felder 11 points learning style measuring scale curtseys by Hawk [17]

Table 6.1 Felder learning style in five dimensions

Preferred learning style dimension	
Sensory-concrete material, more practical, stnd. procedure	Perception
Intuitive-abstract material, innovative, challenges	
Visual learning from pictures, video, multimedia	Input
Verbal-learning from words	
Inductive- It has proved that Eng. students are Inductive	Organization
Deductive	
Active learning by doing group work	Processing
Reflective learning by thinking work alone	
Sequential-learns in step by step, uses broad or partial knowledge	Understanding
Global learning in large jumps, need big picture or big idea	

4. How does the learner priorities the processed information?
   Actively through engagement in physical activity or discussion, or **Reflectively** through introspection.
5. How did the learner progress towards understanding?
   Sequentially in continual steps, or **globally** in large jumps, holistically.

FSLSM is categorized mainly into four dimensions, namely.
(1) Active or Reflective, (2) Sensing or Intuitive, (3) Visual or Verbal, and (4) Sequential or Global [12]. Note: It is a proven fact that engineering students were inductive so the fifth dimension can be ignored.

6.3.2 Learner Preferences in Five Dimensions

Each learner shows specific preferences along each of the four dimensions [12], 18. The intensity of the preferences varied from weak to strong, and thus, the different sets of LS combinations are formed for identification of individual's core LS.

Preferred Learning Styles along with input variables are given dimension-wise in Table 6.1.

6.3.3 Index of Learning Style (ILS) Scale

• 44 questions, 11 for each LS dimensions (Fig. 6.3).

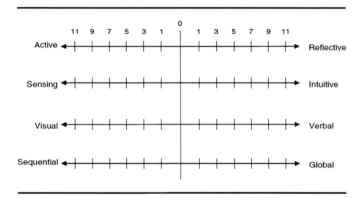

Fig. 6.3 Felder 11 points scale of all four dimensions curtsey by Hawk [17]

6.3.4 Combination of Learning Style [19]

The learners show distinct LSs in combination of 16 groups in one dimension of FSLSM. Each one of them exhibits the prominent characteristics for a learner as undernoted:

1. active + sensing + visual + sequential
2. active + sensing + visual + global
3. active + sensing + verbal + sequential
4. active + sensing + verbal + global
5. active + intuitive + visual + sequential
6. active + intuitive + visual + global
7. active + intuitive + verbal + sequential
8. active + intuitive + verbal + global
9. reflective + sensing + visual + sequential
10. reflective + sensing + visual + global
11. reflective + sensing + verbal + sequential
12. reflective + sensing + verbal + global
13. reflective + intuitive + visual + sequential
14. reflective + intuitive + visual + global
15. reflective + intuitive + verbal + sequential
16. reflective + intuitive + verbal + global

In the framework, FSLSM is applied as a base and guiding model and its learning styles in all four dimensions are taken into considerations. The characteristics of learning activities can be taken from studies [19–23] which are mapped to the learning styles in all four dimensions as shown in Tables 6.2, 6.3, 6.4 and 6.5.

Table 6.2 LS- to LO-oriented activities mapping for dimension 1

Active	Reflective
• Chat, forum posting • SA test (liking MCQs) • Exercises(View /Guessing) • Application/Experimentation • Working actively with learning materials (trying things out) • High communication with others • Working in group and discuss things	• Outline/Concept/Theory/Content Viewing/Summary • Examples • Case studies • Forum viewing • Using online help • Slide show • Result pages view • Think it through, • Work alone

Table 6.3 LS- to LO-oriented activities mapping for dimension 2

Sensing	Institutive
• Interested in viewing examples • Interested in doing exercises • Do SA/quiz more • Do application/experimentation • Go in depth to search in detail about questions • Hand-on experiences/interested in practical materials • Observe slideshows patiently • Case studies • Co-relate with real world • To learn from facts and concrete material • Solve problem with standard approach	• Content viewing • Questions about concepts • Abstract Material/Concepts/Theories/Definition • Conceptual maps • Algorithms • Understand implicit and underline meanings • Able to discover possibilities and established relationship • More creative and innovative than sensing

Table 6.4 LS- to LO-oriented activities mapping for dimension 3

Visual	Verbal
• Graphs and graphics • Chart, tables, and flowchart • Photograph/Images/Picture • Documentary/Demonstrations/Videos • Mathematical equations and formula • Power point slide show and multimedia • Animations/Simulation	• Text-based material • Audio objects • Lesson objectives and content objects • Text slideshows with audio • Difficulty with visual styles • View and post in chat/Forum

6.3.5 Annotating Learning Objects

Each LO relates to a subtype in the set of 16 types of combination shown in Table 6.5.

Table 6.5 LS-to LO-oriented activities mapping for dimension 4

Sequential	Global
• In depth analysis of questions/Interested in detail • Step-by-step exercises • Leaning in order • Pages with few links • View self-assessment test • Use hypertext/Cover hypertext	• Outline of Summary/Lecture/Session/Content • Slide shows • Case studies • Forum viewing • Using online help • Examples • Result pages view • Learn randomly, suddenly found a big picture

6.4 Learner Modeling Based on Fuzzy Cognitive Map (FCM)

The formation of learner models can play pivotal role in any adaptive learning environments [24]. The objective of learner model is to enunciate personalization on the basis of learner's traits and characteristics shown during the learning process. Learning process depends upon cognitive, affective, and behavioral variables related to man–machine interactions. It is aimed to map between learners' actions in the system and the best-suited learning style using FCM techniques for developing a learner model.

6.4.1 Diagnose Learner Profile

In the model, web-based learning and new techno initiatives are introduced to gauge learners' capabilities, their needs, and their preferences during learning process and exploited the information summed up optimally. The uses of FCMs are added to get advantage of analyzing human-related data and finding patterns in adaptive e-learning. An intelligent system can be expressed, deduced, and drawn meaningful conclusions about the characteristics of the learners. The Learning Style Diagnostic Module (LSDM) is a major component of the learner model. Learning styles are meant for different ways of thinking, perceiving, approaching, interacting, and solving problems by the learners. They work efficiently and exploit the limited resources optimally. They prominently exhibit characteristics during the learning process which help to identify LSs. Therefore, the primary use of learning styles is to get ideas for thinking about individual differences and understanding the distinct characteristics. The detection of individualization in the purview of educational technologies is the main concern. There were many a good studies and attempted good research work in the areas of learning styles classifications [17], 25. However, FCMs give better results because of the fact that human analysis in soft knowledge domain is closer to the nature of fuzzification. It seldom requires concrete data and compactness. We use FCMs for better analysis of human-related adaptation in e-learning system.

6.4.2 Collecting and Processing Information

6.4.2.1 Categories of Log Data

- Knowledge Data-Related to subject knowledge, results of pretest, post-test, marks, grade obtain in examination, performance level, etc.
- Chronometric Data-Time spent to read, time for find correct answer, idle time interval, total time on task, time need to review, etc.
- Try Data-Number of attempt to find the correct solution, etc.
- Routing/Navigation Data- Number of times direction finding or course identification over a LO, chapter, session, topic, activity tool or exercise selected, jumping to other topics without attending previous ones.
- Observable and measurable responses are paraphrased into K groups, i.e. in four dimensions of FSLSM.

6.4.2.2 Factors for Detecting Learning Style-Variables and Parameters

- Log **data (Navigational Indicators)**
- Time spent on various learning objects **(Temporal Indicators)**
- Frequency of accessing a particular type of resources, total learner attempts on exercise, assignment, test, etc. **(Performance Indicators)**
- Content-related variables (c-type, c-stay, c-visit)
- Example-related variables (e-visit, e-stay)
- Quiz-related variables (q-stay, q-visit, q-revision, q-ans-change)
- Exercise-related variables (ex-visit, ex-stay, ex-gr)
- Chat-related variables (no-msg, t-chat-listen, t-interactivity), etc.

These variables are useful for mapping learning style to learning dimensions based on FSLSM.

Here, we can measure all measurable learning objects like theory, example, exercise, audio, video, and assignments, whereas some input data and activity variables in terms of learning objects could not be mapped like touch, feel, taste, and smell.

6.4.3   Input Layer

It can be represented by the activities carried out as input variable in the system. We present the usage of processing neurons over the measurable and captured learning activities in the input layer of the network. Some activities in this layer are given as below:

- Interested in particular reading materials: The chapters, sessions, or topics can be placed in abstract forms comprising theories, etc., as well as concrete materials comprising exercises, etc. It can be detected by the system that which type of learning materials are preferred by the learner.
- Interested in accessing to examples: Number of examples can be accessed by the learner from the set of given examples which can be observed for detecting required LS.
- Activities of answer changes: It can reveal hidden characteristic of the learner when the learner changes the answer before end of the examination period. The percentage of such changes can provide important clues about the learner's LS.
- Solving exercises and spending time: Numbers of exercises can be accessed by the learner from the set of given exercises, and how long they stay tuned in time spent can be observed by detecting desirable LS.
- Examination Duration: Time is very important factor in any SA / Test / Examination. It measures efficiency of the learner that how one learner can score better and what is the actual examination delivery time, how many questions are solved, and how one managed the time optimally.
- Revision before Examination or Time in Checking Answers: It is a simple fact that revision before examination enhances marks. During checking correctness of answers, we can understand the facts. Time analysis reveals some important patterns.
- Chat/Forum/Mail/Facebook/Twitter/WhatApps usage: Number of posts ignored/seen/replied/posted, and time spent can provide important database to analyze the patterns for detecting specific learning style.
- Information Access Style: Every learner is different. Each accessed information differently. Some can access learning in sequential, and some can do randomly. The sequential learner learns things in logical and analytical way. The order can be measured, and patterns can be analysed (Table 6.7).

6.4.4   Output Layer

The role of output layer of the network is to provide fairly rough idea about the learning style of learners according to their activities in the input layer. It works as an estimation or approximation tool. The study revealed that how many processing neurons in output layers can be used in each of learning style dimensions of FSLSM.

The processing neurons can also use to subtypes of the dimensions for analyzing further. They are as under noted:

- Perception: It measures the style of the learner in terms of intuitive or sensitive.
- Processing: It decides learner's leaning style in terms of active or reflective.
- Understanding: It informs the learner's LS in terms of sequential or global.
- Input: It identifies the style of the learner in view of visual or verbal.

6.5 Labeling Learning Objects

Labeling of LOs comprising the sequence of learning activities is mapped with subtypes of dimensions of FSLSM as given in Sect. 6.3. The theoretical aspects can be analyzed taking cues of Felder et al. [11], and practical aspects are applied on the note of research study of Popescu et al. [26] and Graf et al. [27] as given in Table 6.6. For example, labeled learning object 1 is mapped as Reflective/sensing/Verbal/Sequential. Similarly, learning object labeled 2 with Visual only.

6.6 Automatic Detection of Learner's Characteristics in E-Learning

In any adaptive e-learning environment, it is imperative to detect learner's individual traits and behaviors which can be shown by the distinct characteristics. They may be very much helpful in finding all about the learning style of the learner. The Behavior Monitoring Module (BMM) gives the platform to collect and corelates all measurable relevant and possible learners' characteristics. The learning activities build numerous log data files which can reveal various information and patterns by applying data mining tools, e.g. number of LOs accessed in a time, number of times a LO was accessed in a time, time spent on a LO in one session, number of postings seen/replied/posted over chat or forum in a time or in a session, number of attempted SA/Test/Quizzes, number of assignments completed in exercises/applications/examples in a time, number of times the assignments attempted, order of the LOs selected by the learner, etc.

6.6.1 Behavior Monitoring Module-BMM

Learner's profile data are characterized in different sets of measurable characteristics in set of S. When learners interact with the system, all the interactive and activity data

Table 6.6 LS- to LO-oriented activities mapping for dimension 4

LSD1(Processing)		LSD2(Perception)		LSD3(Input)		LSD4(Understanding)	
Active	Reflexive	Sensing	Intuitive	Visual	Verbal	Sequential	Global
SA Tests, Exercises, Multiple Type, Questions, Chat, Experimentation	Examples, Outlines, Summaries, Result Pages, Case Studies	Examples, Explanation, Facts, Practical Material	Definitions Algorithms, Abstract Material	Images, Graphics, Charts, Animations, Videos	Text, Audio	Step-by-step, Exercises, Constrict, Link Pages, SA	Outlines, Summaries, All-link, Pages, Example

Table 6.7 Summarizes the input vector, X, representation

X Input Vector	Learning Objects/Activity/Actions	W_{ij} Range $[-1\ 1]$	
x_0	Examples	Low	High
x_1	Revision before test	Less	Much
x_2	Access Facebook/Twitter/WhatApps/Chat/Forum	Limited use or no use	Posted/Seen/Replied /
x_3	Information access	Linear	Global
x_4	Examination delivery time	Quick	Slow
x_5	Exercises	Few	Much
x_6	Reading material	Concrete-much	Abstract-high
x_7	Experiment	Moderate	High
x_n	L_n	–	–

can be automatically captured and stored by the system in terms of log file. During interaction, learner interacts with different learning objects of set $\{L_1, L_2 \dots L_n\}$.

The input data are in terms of number of visits, time period, and order of visiting the learning material.

6.6.2 Auto-Diagnosis of Learner's LS in E-Learning

Learning Style Diagnostic Module (LSDM) a component of learner model identifies a method for detection of learning style already discussed in this section.

However, learner behavioral data which are stored in server database gives the patterns. Learning style can deduce learner behavioral patterns. These patterns are formed by analyzing learner activities data, profile data, and interactive data generated when learner has accessed the learning objects. FCM can assist in mapping the learning style to learning object for each dimension separately [28, 29]. It supports to find the casual relationship among behavioral data which are used as different concepts shown in Figs. 6.4, 6.6, and 6.8. The dependency graph of the model with few parameters and its matrix value for processing dimension and for all dimensions are given in Figs. 6.5, 6.7, and 6.9.

The concept of neural network can be used in the FCM to evaluate the concept value of learning object for a given time and measured the changes in the concept value according to the variables of time. We can apply more general formulation of neural network which is given below:

A more general formulation:

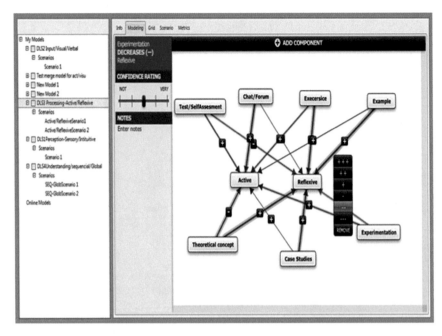

Fig. 6.4 Knowledge mapping using mental modeler for processing dimension

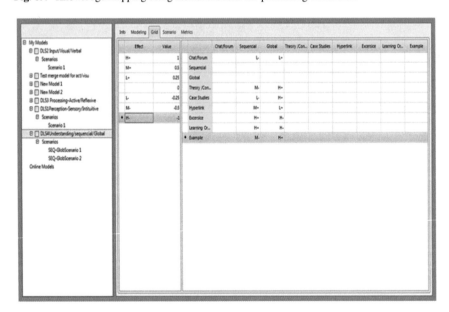

Fig. 6.5 Matrix representation of concept value for processing dimension

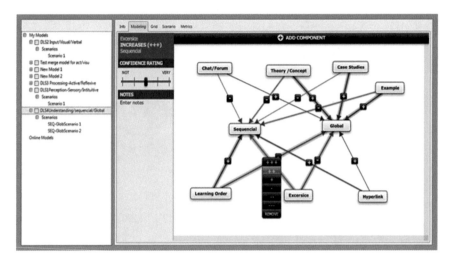

Fig. 6.6 Knowledge mapping using mental modeler for understanding dimension

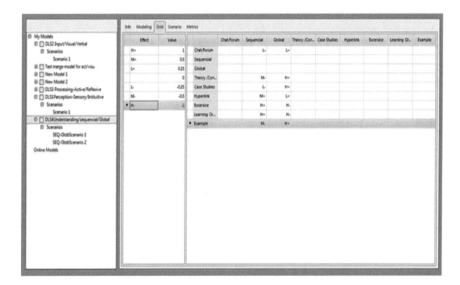

Fig. 6.7 Matrix representation of concept value for understanding dimension

$$P^{n+1}(C_i) = f\left[k_1 \sum_{\substack{j=1 \\ i \neq j}}^{41} w_{ji} P^n(C_j) + k_2 P^n(C_i) \right] \qquad (6.7)$$

where $0 < k_1 < 1$ and $0 < = k_2 < = 1$.

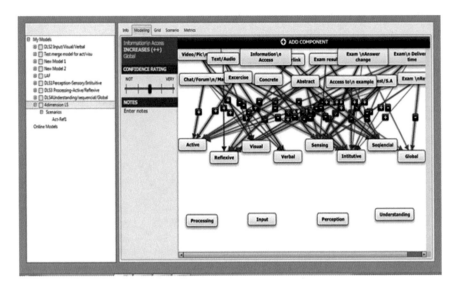

Fig. 6.8 Knowledge mapping using mental modeler for all four dimensions

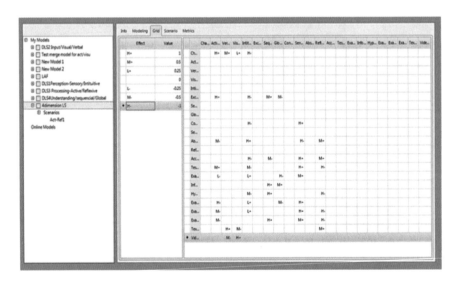

Fig. 6.9 Matrix representation of concept value for all four dimensions

P Value of the concept C.

k_1 the concept's dependence of on its interconnected concepts,

k_2 parameter instrumental for getting new concept value based on previous one.

f a sigmoid function and it was evaluated from $f = \frac{1}{1+e^{-\lambda x}}$.

We have selected $k_1 = k_2 = 0.5$ as the results vary slightly when actual values are taken after each recalculation. It is closer to the discrete final values. Other squeezing functions tanh(x), tan(x), etc., applied to change the results into the fuzzy range [0, 1] or [−1, 1], where concepts took new values in between.

Algorithm has been set for the input data based on unipolar sigmoid function, and Sugeno fuzzy inference method (IF–THEN rule) has been applied to obtain patterns of output data. The obtained concept values are fuzzified in terms of linguistic variables (low, medium, and high). The strengths between learning objects measured in crisp values are also fuzzified in terms of linguistic variable. The other important parameters like motivation and emotions can be added to fine tune the personalized adaptivity process and strengthen the model. These parameters can also be obtained from the activities of the learners while working with the e-learning objects/course materials. Analysis of data and cross-validation with the results obtained by the research scholars can be minutely examined and can be tallied, and the precision values can be found.

There are many prominent data mining techniques used by the researchers to detect the underlying patterns from a large pool of data. Most of the prevalent mapping system use crisp approach, i.e., only binary value either 0 or 1 to understand the concepts and mapping inferences. Crisp logic does not precisely represent human understanding as it comes under the domain of soft knowledge computing. Therefore, FCM can be used in such cases as human aspects are related to the soft skills. Fuzzy logics can be represented in terms of truth values on a continuous scale from 0 to 1 (neither wholly true, nor wholly false). It is defined in a range. Moreover, FCM is a combination of fuzzy logic and neural networks. It is a concept of mapping various relevant parameters and their interconnected causal relationship. In Figs. 6.4, 6.6, and 6.8, the concepts are denoted by node C_i and the edge denoted by w_{ij} which weight measured the strengths of one concept on another. Its value varies from [−1 1]. The value of concept and the interconnected weights is measured by neural network and Hibbean rule as mentioned above in mathematical analysis of FCM. The value of concept and interconnected weight of strength may vary with time. Therefore, the concept values are dynamic and so the learning styles and adaptive e-learning.

FCM is useful for creating meta-knowledge and exploring hidden implication of learner's understandings. The cognitive map pictorially exhibits interrelationship among various learning objects as well as learning ability factors. It is a presentation of the perception and beliefs of a decision maker or expert system about own subjective world.

For simplicity, fuzzy variables are taken into three degrees of linguistic variable (low, medium, and high) for measurement of learning styles. Threshold membership values for the classification of linguistic variables are taken as per the literature based on [20]. Range varied from [−1 1] positive term defines one pole of each learning style as follows:

(a) 0 to 0–0.3 weak, (b) 0.3 to 0.7-moderate, (c) 0.7–1.0 strong.

6.7 Measurements of Concepts Values and Their Strengthens Casual Relationship

Let us S is a set of Student's.

$$S = \{s_1, \ s_2, \ s_3 \ldots s_k\} \ \mathrm{k} \ > \ 0 \ \text{for numeral} \tag{6.9}$$

where k = number of students.

Let us T be set of data capturing tool/activity measuring variables.

$$T = \{t_1, \ t_2, \ t_p\} \ (t_1 = \text{time related on Lo}, \ t_2 = \text{frequency related data}) \tag{6.10}$$

where t $_p$ (independent set) = activity measuring variables like browsing pattern, participant's log in/log out data, visiting course materials, skip, number of attempts for any activity/LO, time consumed, speed action, and learning order.

Let L be the set of learning objects/activities.

$$L = \{L_1, \ L_2, \ L_3 \ldots L_n\} \tag{6.11}$$

where L_n = measurable learning objects in domain of e-learning defined in Tables 6.2, 6.3, 6.4, and 6.5 like headline /Overview, summary, Abstract material, theories, concrete material, definition, example, hypertext, Graph/picture, animation, illustrations, inferences, simulation, video, audio/text, ppt, mail, chat/forum social networking platforms: Twitter, Facebook, WhatsApp, interpersonal, group work, experiment, application, assignment, case studies, scenario analysis, expanding, test, self-assessment, examination revision, view self-assessment test, pre-test, post-test, quiz, course completion, feedback, etc., can be taken as concepts. Each and every concept has a certain value and as the time passes through the values of the concepts change gradually as they are under the impact area of the nearest concepts and related weights in between the concepts.

Let b be the set of concept values of learning objects up to nth terms in terms of number or frequency, time, order, and speed, i.e. activity measuring variables as per set T.

$$b = \{f(L_1, t_1), \ f(L_1, t_2) \ldots f(L_2, t_1), \ f(L_2, t_2 \ldots .f(L_2, t_p) \ldots .$$
$$f(L_3, t_1), \ f(L_3, t_2) \ldots .f(L_1, t_p)\} \tag{6.12}$$

For independent variable $f \ (L_i, \ t_i) = 0$, where correlation not established. For example, order does not matter in exercise, video, picture, graph, etc.

Fuzzification of values of learning objects.

$$B_{ij} = \{f(b_{11}), \ f(b_{12}) \ldots .f(b_{21}) \ldots ..\} \tag{6.13}$$

For each b_{ij} ($i = 1, 2.... 1$ & $j = 1, 2, 3......$p).
$B = \{b_1, b_2......b_n\}$ it is written as b_{ijk}^k
The concept value of learning objects is represented in terms of concept.
C_i Eq. (6.5) and can be converted into Eq. (6.6) as given below for calculating new changed value. Let at the step $n + 1$, the value $P^{n+1}(C_i)$ of the concept C_i calculated by the equation of neural network:

$$P^{n+1}(C_i) = f\left[\sum_{\substack{j=0 \\ j\neq 0}} w_{ij} P^n(C_j)\right] \qquad (6.14)$$

where $P^n(C_i) = $ value of the concept C_i at the discrete time step n.
We applied a more general formulation in the proposed study:

$$P^{n+1}(C_i) = f\left[k1 \sum_{\substack{j=1 \\ i\neq j}}^{4l} w_{ji} P^n(C_j) + k_2 P^n(C_i)\right] \qquad (6.15)$$

where $0 < k1 < 1$ and $0 < = k2 < = 1$.

k_1 the concept's dependence of on its interconnected concepts,
k_2 the proportion of contribution of the previous value of the concept in the computation of the new value.
f a sigmoid function, $f = \frac{1}{1+e^{-\lambda x}}$

We select $k_1 = K_2 = 0.5$ as the results are in miner variation of the values of the concepts after each recalculation and closer to the discrete final values. Other squeezing functions $\tanh(x)$, $\tan(x)$, etc., can be applied to change the results into the fuzzy interval $[0, 1]$ or $[-1, 1]$, where concepts took new values.

At each time step, value of C has changed for new value recalculated as per above equation [30], 7. For one group of value of C after normalizing value from $[0, 1]$, it can be fuzzified in following terms:

where if $A = 2$, i.e. {low, high}.

if A 3, i.e., {low, medium, high}.
if A 5, i.e., {very low, Medium, High, Very high}.

For calculating weight between two concepts, we use differential Hebbian learning law [31] which is represented by:

$$W_{ij} = w_{ji} + x_i x_j \qquad (6.16)$$

So $x_i x_j > 0$, if C_i and C_j moved in the same directions and $x_i x_j < 0$ if C_i and C_j moved in opposite directions. Therefore, concepts which tended to in the same line

Table 6.8 Membership value in each learning style of a learner

Dimension 1		Dimension 2		Dimension 3		Dimension 4	
ACT	REF	SEN	INT	SEQ	GLO	VIS	VER
0.79	0.11	0.29	0.75	0.86	0.28	0.75	0.24

Table 6.9 Classification of LSs based on strength

Dimension 1		Dimension 2		Dimension 3		Dimension 4	
ACT	REF	SEN	INT	SEQ	GLO	VIS	VER
S	W	W	S	S	W	S	W

have strong positive weights, while those that tended to be in opposite has strong negative weights [7].

Learning styles can be obtained by incorporating fuzzy rules in FCM techniques. **Mapping of Sets into Concepts and Normalized it into [0 1]**.

To define these methods, the following linked approaches are necessary to map according to the sets.

1. Set of elements $C_i \in \Theta$, where $\Theta = \{B\}$.
2. A, linguistic variable.
3. A calculated numeral assignments in the interval $X \in (-1, 1)$.
4. $P \in X$, a linguistic variable $C_i \in \Theta$.
1. $5 \mu A (C_i)$ was obtained the degree of membership of θ_i to the set of elements calculated by A (Tables 6.8 and 6.9).

6.8 Fuzzy Rule Base in Terms of Adaptive Rules

Some basics samples (few rules) adaptive rules can be generated and described as:

1. If (num-of-chat-Post is High) and (num-of-SA is High) and (t-spent-on is Low) and (num-of-exercise is High) and (No_of_experiment is High), then (Processing(ACT/REF) is Sact) (1).
2. If (num-of-chat-Post is Medium) and (num-of-SA is Medium) and (t-spent-on is Low) and (num-of-exercise is Medium) and (No_of_experiment is Medium), then (Processing(ACT/REF) is Mact) (1).
3. If (num-of-chat-Post is Medium) and (num-of-SA is Low) and (t-spent-on is Low) and (num-of-exercise is Medium) and (No_of_experiment is LOw), then (Processing(ACT/REF) is Wact) (1).
4. If (num-of-chat-Post is Medium) and (num-of-SA is Low) and (t-spent-on is High) and (num-of-example is High) and (num-visits-abstract is High) and (Case_study is High), then (Processing(ACT/REF) is Sref) (1).

5. If (num-of-chat-Post is Medium) and (num-of-SA is Low) and (t-spent-on is High) and (num-of-example is Medium) and (num-visits-abstract is Medium) and (Case_study is Medium), then (Processing(ACT/REF) is Mref) (1).

References

1. Li S-J, Shen R-M (2004) Fuzzy cognitive map learning based on improved nonlinear Hebbian rule. Processing 2004 international conference machine learning cybernetics (IEEE Cat. No.04EX826) vol 4. pp 256–268. https://doi.org/10.1109/ICMLC.2004.1382183
2. Cole JR, Persichitte KA (2000) Fuzzy cognitive mapping: applications in education. Int J Intell Syst 15:1–25. https://doi.org/10.1002/(SICI)1098-111X(200001)15:1<1::AID-INT1>3. 0.CO;2-V
3. Papageorgiou EI, Salmeron JL (2013) A review of fuzzy cognitive maps research during the last decade. IEEE Trans Fuzzy Syst 21(1):66–79. https://doi.org/10.1109/TFUZZ.2012.2201727
4. Georgiou DA, Botsios S, Mitropoulou V, Papaioannou M, Schizas C, Tsoulouhas G (2011) Learning style style recognition recognition based based learning on on three-layer adjustable three-layer fuzzy cognitive map. 1(4):333–347
5. Changing T, Scene A (2004) Introduction 1.1. pp. 1–48. https://doi.org/10.1016/B978-0-444-53604-4.00001-6
6. Mago V, Fellow P (2011) Fuzzy logic and fuzzy cognitive map
7. Stylios C, Groumpos P (1999) Mathematical formulation of fuzzy cognitive maps. Processing 7th Mediterranean …, pp 2251–2261. https://www.researchgate.net/publication/228567505_Mathematical_formulation_of_fuzzy_cognitive_maps/file/d912f508a55f2b6fb4.pdf
8. Margaritis M, Stylios C, Groumpos P Fuzzy cognitive map software
9. Sweta S, Lal K (2017) Personalized adaptive learner model in e-learning system using FCM and fuzzy inference system. Int J Fuzzy Syst 19(4):1249–1260. https://doi.org/10.1007/s40815-017-0309-y
10. Ross TJ (2004) Fuzzy logic with engineering applications
11. Designs E (2005) Understanding student differences. 94(1):57–72
12. Felder R, Silverman L (1988) Learning and teaching styles in engineering education. Eng Educ 78(June):674–681. https://doi.org/10.1109/FIE.2008.4720326
13. Gaikwad T, Potey MA (2013) Personalized course retrieval using literature based method in e-learning system. Processing—2013 IEEE 5th International Conference Technology Education T4E 2013. pp 147–150. https://doi.org/10.1109/T4E.2013.44
14. Zywno MS (2003) A contribution to validation of score meaning for Felder-Soloman's index of learning styles. Engineering Education pp 1–16. https://citeseerx.ist.psu.edu/viewdoc/download?doi=10.1.1.167.7813&rep=rep1&type=pdf
15. Nam V (2013) A method for detection of learning styles in learning management systems. 75
16. Carmona C, Castillo G, Millan E (2008) Designing a dynamic bayesian network for modeling students' learning styles. 2008 Eighth IEEE Int Conf Adv Learn Technolhttps://doi.org/10.1109/ICALT.2008.116
17. Hawk TF (2007) To enhance student learning. Decis Sci J Innov Educ 5(1):1–19
18. Premlatha KR, Geetha TV (2015) Learning content design and learner adaptation for adaptive e-learning environment: a survey. Artif Intell Revhttps://doi.org/10.1007/s10462-015-9432-z
19. Dung PQ, Florea AM (2012) An approach for detecting learning styles in learning management systems based on learners ' behaviours. 2012 Int Conf Educ Manag Innov 30:171–177
20. Yang J, Huang ZX, Gao YX, Liu HT (2014) Based on a pattern recognition technique. 7(2):165–177. https://doi.org/10.1109/Tlt.2014.2307858
21. Saberi N, Montazer G (2012) A new approach for learners' modeling in e-learning environment using LMS logs analysis. 2012 Third international conference e-learning e-teaching. pp 25–33

22. Kusumawardani SS, Prakoso RS, Santosa PI (2014) Using ontology for providing content recommendation based on learning styles inside e-learning. 2014 2nd international conference artifical intelligence modelling simulation pp 276–281. https://doi.org/10.1109/AIMS.2014.40

23. Franzoni AL, Assar S (2009) Student learning styles adaptation method based on teaching strategies and electronic media. Educ Technol Soc 12(15–29)s. https://ieeexplore.ieee.org/xpls/abs_all.jsp?arnumber=4561832

24. Villaverde JEÃ, Godoy DÃ, Amandi AÃ (2006) Learning styles ' recognition in e-learning environments with feed-forward neural networks. pp 197–206

25. Kolb DA (1981) Learning styles and disciplinary differences. Reponding to the new realities of diverse students and a changing society. 232–255. https://doi.org/10.1016/S0002-8223(97)00469-0

26. Popescu E (2010) Adaptation provisioning with respect to learning styles in a Web-based educational system: an experimental study. J Comput Assist Learn 26(4):243–257. https://doi.org/10.1111/j.1365-2729.2010.00364.x

27. Kinshuk SG, Liu TC (2008) Identifying learning styles in learning management systems by using indications from students' behaviour. Processing—8th IEEE international conference advanced learning technologies ICALT pp. 482–486. https://doi.org/10.1109/ICALT.2008.84

28. Gray SA, Gray S, Cox LJ, Henly-Shepard S (2013) Mental modeler: a fuzzy-logic cognitive mapping modeling tool for adaptive environmental management. Proc Annu Hawaii Int Conf Syst Sci 965–973. https://doi.org/10.1109/HICSS.2013.399

29. Gray S An introduction to MENTAL MODELER: a tool for environmental planning and research

30. Georgiou D, Botsios S, Tsoulouhas G, Karakos A Adjustable learning style recognition based on 3layers fuzzy cognitive map

31. Georgiou DA, Makry D (2004) A learner's style and profile recognition via fuzzy cognitive Map. Processing—IEEE international conference advanced learning technologies ICALT. pp 36–40. https://doi.org/10.1109/ICALT.2004.1357370

Chapter 7
Recommender System to Enhancing Efficacy of E-Learning System

7.1 Introduction

Personalization systems record and utilize every bit of valuable information about each of the learners. In the system, learners' behaviours influence their preferences which are the ingredients of the core concepts in terms of making right decisions while recommending and providing learning materials [1]. The system presumes what types of personalized materials or services are required to deliver based on learner's choices and expectations. The choices and expectations are data collected from either from learners' profile or captured during man–machine interactions. It is either be expressed or hidden. It is revealed that each individual learner has different mindset, think differently, react randomly, and show dissimilar personal traits. It is very challenging to provide personalized materials and to arrange matching adaptations. The comprehensive study of cognitive learning is included many more components difficult to creating different clusters accordingly and measure, but they can affect the learning process. It is important that a small part of inferences of these components can be measured and learning preferences can be indicated to some extent. They are motivation, emotion, sentiments, and other knowledge ability factors.

As per García Barrios [2], personalization was a part of adaptation linked with learner's traits, whereas adaptation referred to the concept of alignment with all entities which were related to the user [3].

We can present some concrete ideas to express and extract core knowledge across the system. It emphasizes to explore a way out for the most suitable system to implement and get the desired adaptive e-learning for the learner.

The potential effectiveness of Adaptive Neuro-Fuzzy Inference System (ANFIS) can be examined (as discussed in previous chapter) applying hybrid learning system. The upgraded system can be used for providing adaptive learning materials as and when learner demands the same.

© The Author(s), under exclusive license to Springer Nature Singapore Pte Ltd. 2021   87
S. Sweta, *Modern Approach to Educational Data Mining and Its Applications*,
SpringerBriefs in Computational Intelligence,
https://doi.org/10.1007/978-981-33-4681-9_7

Outcomes of the experiments can be shown that ANFIS model has worked successfully in adapting the learning materials in the respective manifestations. The results can be arrived by using standard error measurements. It is found that the optimal setting is very much required to forecast and that is in a better way. The results of the MATLAB-enabled simulations show that the ANFIS approach can be more worthy and simple to run the process. They can be deduced from the study of different models in which different input variables, membership functions, sample size, etc., are applied. However, it is important that more input variables tend more response time for the e-learner.

7.2   Personalized Adaptive Module-PAM

This module is set to deliver personalized learning materials based on some core dependency factors. For simplicity, author has taken as just three factors—learning style (LS), motivation (M), and knowledge level (KL). This method basically incorporates two cognitive factors—motivation and knowledge ability—with learning style and learning preferences can be identified for a learner. Qualitative parameters like motivation and knowledge can be estimated for better understanding of learner preferences.

7.2.1   Calculating Motivation (High/Low)

We have taken apt educational strategies and can put some threshold ideas which can invite due attentions of the learners to participate vigorously in the learning process. They can provide platform for smooth discussions across the board, improvise assignments, set simple to complex quizzes, take feedbacks intended or unintended, exploit maximum use of multimedia, etc. Motivation can be estimated on the bases of data available in log files in terms of learning objects chat, forum, blog, and more learning in less time. It can be either low or high. The strength of log data and their patterns can clearly show the motivation level. The calculated values can be fuzzified, and threshold limits in form of two membership functions can be set to determine motivation either low (below 50%) or high (above 50%). Inclusion of these motivation factors may enhance e-learning process substantially.

7.2.2   Calculating Knowledge Ability (Low/Medium/High)

The knowledge ability factors (KNF) deduced from the above-mentioned tools can be used for enhanced e-learning adaptivity which enhanced overall learning process. It is a measurement of prior knowledge level, and the system can understand to deliver

Table 7.1 Combinations of learning preferences inclusive of three factors

Learning style	Motivation (MT1-low, MT2-high)	Knowledge ability (KNA1-Low, KNA2-Medium, KNA3-High
LS1	MT1	KNA1
	MT1	KNA2
	MT1	KNA3
	MT2	KNA1
	MT2	KNA2
	MT3	KNA3

the contents as important factor into learning. It enhances e-learning performances because of added adaptation by using the additional cognitive factor. The system can examine the knowledge ability, by test, self-assessment test, quizzes, online debates, etc. in terms of: Low (0–30%) / Medium (31–74%) / High (75–100%) can be represented by three membership functions.

$$LS = \{LS1/LS2/LS3/.../LS16\} \tag{7.1}$$

$$ThenMT = \{Low(MT1)/Medium(MT2)\} \tag{7.2}$$

$$AndKNA = \{(Low(KNA1)/Medium(KNA2)/High(KNA3)\} \tag{7.3}$$

$$LSP = \{\{LS\}U\{MT\}U\{KNA\}\} \tag{7.4}$$

In Table 7.1, we can show how learning objects combined in set (C1...C40), which is formed by accumulating above factors, such as: C1 = a learning style preference active/sensing/visual/sequential (LS1), with motivation high (MT2) and ability of knowledge being Low (KNA1).

In this chapter, learner preferences can be calculated in combination of matching learning style, motivation, and knowledge factor. More learning ability factors can be taken into consideration. The possible preferences only for one learning style dimension is given in the tabular form in Table 7.1 as under noted. A total of 96 learning style preferences can be formed.

LSP = {LSP1, LSP2....LSPn} where n = 96.

LSP1 = {LS, MT, KNA} e.g. LSP1 = {LS1, MT2, KNA1).

Each learner's data is formed by all three factors—learning style (LS), motivation (MT), and knowledge ability (KNA).

In this module, learner gets personalized adaptive learning materials as per their identified learning preferences. Scientifically, it is proved that learner's results or outcome can be enhanced by more than 26% in adaptive learning system [4].

7.3 Overview of Implementation of a Novel Framework (NAPF)

Fuzzy Inference System (FIS) and Adapted Neuro-Fuzzy Inference System (ANFIS) models can be used to implement the e-learning environment in detection of learning style preference (LSP) to offer personalized learning materials as demanded by the learner. It combines the applications of neural network and fuzzy logics. The ANFIS model can be developed based on various experiments with different membership functions (MFs) used, selection of different numbers of epoch, etc. They can be segregated in accordance with five ANFIS modules to understand the findings of different learning activities and processes in the given scenarios. MATLAB-enabled computer simulation can be performed for each one of the scenarios, and the result can be validated accordingly to select the best model using the indexes of statistical analysis.

The fuzzy system is mainly a knowledge base system which comprises of IF–THEN rules [5]. Fuzzy logics are simple way of representing imprecise data. It is combined with neural networks to produce self-learning algorithms, better known as Neuro-Fuzzy systems. Different types of membership functions can be used, e.g. triangular, trapezoidal, generalized bell shaped, Gaussian curves, polynomial curves, and sigmoid functions.

7.4 Fuzzy Inference Systems (FIS)

A Fuzzy Inference System (FIS) is basically function as mapping input data vector into nonlinear into a scalar output using fuzzy rules [6, 7]. The components of FIS are shown in Fig. 7.1, and the procedure has following steps as shown in Fig. 7.2.

- Input variable/Crisp output
- Membership functions or degrees of membership
- Fuzzy logic operators to derive results
- Fuzzy logics like if–then rules

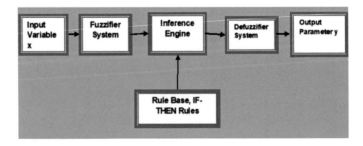

Fig. 7.1 Block diagram of fuzzy inference system

Fig. 7.2 Fuzzy logic control systems

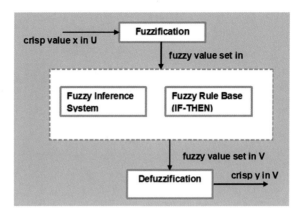

- Aggregation of output in linguistic sets, and
- Defuzzification from linguistic to crisp values.

7.4.1 Sugeno Fuzzy Inference Method

In a Sugeno model [8], each rule has a crisp output generated by a defined function. Therefore, the overall output is evaluated via a weighted average defuzzification, as shown in Fig. 7.3.

The fuzzy rules in the ANFIS model may be presented in the following form:

Rule 1: IF $(x_1$ is $A_1)$ AND $(x_2$ is $B_1)$ AND $(x_3$ is $C_1)$, THEN $(f_1 = a_1x_1 + b_1x_2 + c_1x_3 + d_1)$.

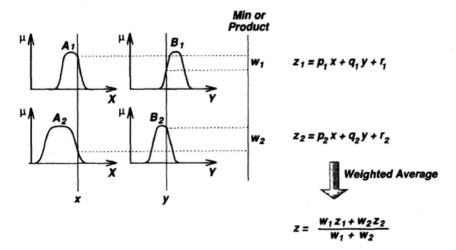

Fig. 7.3 Sugeno Fuzzy Inference Model with two-rule courtesy by [9]

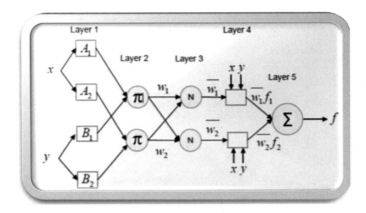

Fig. 7.4 ANFIS architecture

Rule 2: IF (x_1 is A_2) AND (x_2 is B_2) AND (x_3 is C_2), THEN ($f_2 = a_2x_1 + b_2x_2 + c_2x_3 + d_2$).

Sugeno fuzzy method is used in Fuzzy Inference System as shown in Fig. 7.3. Table 6.9 in Chap. 6 shows an example of membership values obtained in each dimension of a student.

7.5   ANFIS Editor and Training

An Adaptive Neuro-Fuzzy Inference System (ANFIS) is a fuzzy system in which variables of membership function can be calculated by introducing neuro-adaptive learning methods in the neural networks [10, 11]. In ANFIS, a hybrid algorithm is used, which is a combination between gradient descent and least squares method. The ANFIS architecture for two rules is given in Fig. 7.4. Circle represents the fixed node, whereas square shape denotes adaptive nodes.

The outcomes of ANFIS can be analyzed by standard error measurement techniques which show that optimization is mandatory for forecasting in a better way. The values of fuzzy ranges taken only from the system of experts are inadequate to derive the required conclusion.

7.6   Model Training

The learning rule of ANFIS training is a combination of gradient descent and the least squares method. It is used in adaptive personalized learning to achieve the best alternative performance. ANFIS training created a set of suitable training data to train the Neuro-Fuzzy. The number of training epochs and the training error tolerance can be obtained to set the stopping parameters for training. The training process stops when it reached the maximum epoch number, or the training error goal can be achieved.

Overfitting is a common problem in this model, when the data are over trained by ANFIS. The optimal number of epochs (40) is only found through experiments. Overfitting can be analyzed by plotting the training and checking error values.

7.7 Validation of Model

Validation of the model can be carried out by testing or checking data after FIS is trained. It is accomplished with the ANFIS Editor GUI using the testing and checking data sets. The checking data set can be used to avoid the overfitting of the data. When the checking and training data are presented to ANFIS, the FIS model's parameters associated with the minimum checking data model error are selected.

References

1. Al- A, Shen J, Member S, Al-hmouz R, Yan J (2012) Modeling and simulation of an adaptive neuro-fuzzy inference system (ANFIS) for mobile learning. IEEE Trans Learn Technol 5(3):226–237
2. Medwell J et al (2019) Concept-based teaching and learning: integration and alignment across IB programmes a report to the international baccalaureate organisation
3. Botsios S, Georgiou D (2008) Recent adaptive e-learning contributions towards a standard ready architecture. e-Learning. https://utopia.duth.gr/~dgeorg/PUBLICATIONS/53.pdf
4. Kavčič A (2004) Fuzzy user modeling for adaptation in educational hypermedia. IEEE Trans Syst Man Cybern Part C Appl Rev 34(4):439–449. https://doi.org/10.1109/TSMCC.2004. 833294
5. Acampora G (2011) A TSK neuro-fuzzy approach for modeling highly dynamic systems. 2011 IEEE Int Conf Fuzzy Syst (FUZZ-IEEE 2011). pp 146–152. https://doi.org/10.1109/FUZZY. 2011.6007638
6. Saxena N, Saxena KK (2010) Fuzzy logic based students performance analysis model for educational institutions. (January):79–86
7. Casillas J, Cordón O, Herrera F, Magdalena L (2013) Accuracy improvements to find the balance interpretability-accuracy in linguistic fuzzy modeling: an overview. pp 3–24. https://doi.org/10.1007/978-3-540-37058-1_1
8. Ross TJ (2004) Fuzzy logic with engineering applications
9. Jang J-S (1993) ANFIS: adaptive-network-based fuzzy inference system. IEEE Trans Syst Man Cybern 23(3):665–685
10. Matlab (2015) Fuzzy logic toolbox functions. https://www.mathworks.com/help/fuzzy/functionlist.html
11. Yusof N, Zin NAM, Yassin NM, Samsuri P (2009) Evaluation of student's performance and learning efficiency based on ANFIS. SoCPaR 2009—Soft Comput Pattern Recognit 460–465. https://doi.org/10.1109/SoCPaR.2009.95

Printed in the United States
By Bookmasters